BROWNING AUTOMATIC RIFLE

FROM THE 1918 TO THE 1918A3 SLR

Paul Ruffin, with Bob Conroy

Texas Review Press
Huntsville, Texas

Printed in the United States of America

FIRST EDITION

Requests for permission to acknowledge material from the work should be sent to:

Permissions
Texas Review Press
English Department
Sam Houston State University
Huntsville, TX 77341-2146

Acknowledgements:

The authors would like to thank Ohio Ordnance Works for the company's assistance in bringing this volume to print, especially Tara Landies and Sarah Yeary for helping provide techinical information and photographs. Thanks to Bob and Robert Landies for their phone interviews and to Texas Review Press intern Keely Disman for helping transcribe the recording of those interviews. Further, we wish to thank the Browning Museum and the Missouri Highway Patrol Museum for the photographs they provided.

Library of Congress Cataloging-in-Publication Data:

Conroy, Bob, 1977- author.

Browning automatic rifle : from the 1918 to the 1918A3-SLR / Bob Conroy and Paul Ruffin. -- Edition: First.

pages cm

ISBN 978-1-937875-81-7 (pbk. : alk. paper)

1. Browning automatic rifle--History. I. Ruffin, Paul, author. II. Title.

UD395.B8C66 2015

623.4'424--dc23

2014045548

In memory of John Moses Browning
and his great contribution to American military firearms:
The Legendary BAR

CONTENTS

I

JOHN MOSES BROWNING: THE LEGEND

II

THE BAR: A LONG, DISTINGUISHED HISTORY

III

EVERY MAN'S BAR: THE OHIO ORDNANCE 1918A3 SLR

Introduction

My introduction to the Browning Automatic Rifle (BAR) was in Advanced Infantry Training at Fort Jackson, South Carolina, when a handful of us recruits were called on to fire what by that time had already become a relic, relegated to National Guard Armories and training facilities like ours. At that time, the country was not engaged in a major war, so the pressure was not on to develop a light machine gun to replace the BAR. Besides, BARs were plentiful and had a long history of reputable performance.

The lightweight M14, essentially a box-magazine-fed Garand with the capability for automatic fire, had already been adopted and issued to forces in the field in the 7.62 NATO caliber, but it in no way measured up to the robust BAR, and the M-60, a belt-fed 7.62 medium machine gun designed to replace both the BAR and Browning 1919, was already in the hands of the active Army.

I instantly fell in love with that old BAR, a 1918A2 that had obviously been on the beat for many years, with the parkerizing well-worn and the wooden foream and Bakelite butstock bearing dings and scratches and gouges sufficient to send a lesser gun back to the armory or to the discard pile. Hey, it accepted a magazine, seated the first round, and fired as long as I held down the trigger, which I did until the magazine was empty. I don't know how many magazines I ran through that beast, but somebody on the line had to pry it out of my hands. I especially liked the low-speed setting (350 rounds a minute or so), the slow, steady hammering of my target with rounds of .30-06. I would drift off to sleep at night recalling the wonderful rhythm of sound.

Heavy, hell yeah, but what that meant to me was something that would stand up to any kind of abuse a soldier might throw at it, including prying a jeep out of a ditch, and still rattle away. I dearly loved that old gun.

Since those days at Fort Jackson I have often thought of my BAR and how it felt, looked, and sounded, and I longed for the day that I might be able to own one of my own. But in the civilian market they were hard to come by, and the costs were high. Besides, what did a high-school English teacher need with a big old fully automatic BAR, even if he found one and his budget could afford it?

So way led on to way, and my BAR slid quietly to back of my mind, along with the M1 Garand and M1 Carbine and Colt 1911.

In time I bought a Garand and carbine and 1911 through the NRA CMP program, but the BAR was just a dream, and it remained so until a few years ago I ran across an ad for a semi-automatic version manufactured by a company in Chardon, Ohio: Ohio Ordnance Works. But there was a waiting list a mile long, and the guns were expensive, so I shelved the notion of buying one. But I didn't forget about it.

In 2013 I finally bought my1918A3 SLR from OOW, and I was so impressed with the job that they had done on it that my passion for the old war horse was rekindled, and I wrote a series of newspaper articles about the weapon, tracing its history from the John Moses Browning prototype to the 1918A3 that I had in hand. Everything about the OOW rifle declared quality engineering, and I had a ball breaking it down and cleaning it after its first firing. Every part was so robust, everything about it so authentic feeling.

In the process of trying to make certain that I had all my facts right about the A3, I struck up an email correspondence with Bob Conroy, Vice President of OOW, and out of our exchanges came the idea for a history of the BAR from its beginning to its latest issue.

This book, then, is an examination of the Browning Automatic Rifle, better known as the BAR, from its conception in the mind of John Moses Browning through its latest reintroduction in three semi-automatic versions by Ohio Ordnance Works. It does not presume to be a comprehensive history of the weapon: If you want that, you need nothing more than James Ballou's *Rock in a Hard Place*, a book that no one interested in the BAR should be without. It is a well-written, thorough, exhaustively researched, well-illustrated treatment of Browning's classic creation.

No book focusing on the BAR can easily avoid an examination of the genius behind the gun, John Moses Browning, so much of the book focuses on the imagination and hard work that led to the introduction of the weapon in World War I. Like a number of Browning's creations, the BAR has stood the test of time and served our country for many decades, indeed, for almost a century. Few weapons over the course of history can boast such a long and illustrious career.

Most of the background of John Moses Browning and the history of the BAR during, between, and after the major wars of the 20th century may be found in myriad sources (in more or less the same wording), as may relevant photographs, so our treatment of that history is cursory with little in the way of documentation. Over the years I have read and absorbed a tremendous amount of knowledge on the BAR and the Brownings and their contributions to the world of firearms, so determining the original sources of much of the material would be impossible anyway. The bibliography at the end of the book is limited in scope and contains no Internet references and no listing of government manuals associated with the BAR. Again, if you are looking for the most comprehensive, well-researched study of the weapon available, then Jim Ballou's *Rock in a Hard Place* is the place to go. If you wish to read the most accessible, thorough, and accurate treatment of the life and accomplishments of John Moses Browning, I would suggest *John M. Browning, American Gunmaker: An Illustrated Biography of the Man and His Guns,* by John M. Browning (John's eldest son) and Curt Gentry.

This book is brief overview of the man of genius behind the Browning Automatic Rifle and a celebration of the weapon from its introduction to the world of American military firearms to its latest variants, concluding with special emphasis on the role of Ohio Ordnance Works in keeping this legendary weapon alive and well and making versions of it available to collectors and shooters around the world.

Few Americans who collect and/or routinely shoot military firearms are willing to spend the money and go through the hassles of purchasing and possessing an original full-auto BAR. It is refreshing to know that a company like Ohio Ordnance Works is devoted to keeping this legend alive.

Paul Ruffin
Texas State University System Regents' Professor
Sam Houston State University

BROWNING AUTOMATIC RIFLE

FROM THE 1918 TO THE 1918A3 SLR

I

JOHN MOSES BROWNING:

THE LEGEND

Johnathan Browning

From Sturdy Stock:

Johnathan Browning

Brushy Fork

It is difficult to imagine what direction the mind of John Moses Browning might have moved in had he not been the son of Johnathan Browning, himself a highly accomplished blacksmith, locksmith, gunsmith, and overall mechanical genius.

The elder Browning, born in 1805 and raised in Brushy Fork, Tennessee, was the son of a hard-scrabble dirt farmer in a county in which one's opportunities for employment were severely limited: He farmed, or he sought employment in the towns and cities, where actual cash was paid for labor.

Brushy Fork, in the northeast corner of the state, was fifty miles from any significant population center and did not even have a school where Johnathan might at least through education have bettered his chances in life.

As is often the case, a fluke occurrence altered a single life in such a way that it might well be regarded as the triggering event for all that followed. Indeed, had it not happened, we might well never have benefited from the great range of weapons produced by the Browning dynasty.

Sometime during his early teens, while doing chores for a neighboring farmer, Johnathan discovered an old flintlock rifle, broken and useless, that the farmer had determined beyond repair and simply junked. In exchange for a week of labor on the farm, the neighbor gave the rifle to young Browning, who carried it home to try his hand at repairing.

Johnathan had been spending a good bit of time at a blacksmith shop not far from his home, mostly observing, so he had picked up some notions of metal working and the way mechanical things functioned. He had a quick and able mind that was especially adept at comprehending mechanical processes. It is not at all difficult to see where John Moses Browning's mechanical acumen came from.

After disassembling the broken flintlock and determining what needed to be done to restore it to the useful and the good, Johnathan made the necessary parts, reassembled the gun, tested it, and proudly took it back to the neighbor, who paid him four dollars for the repair.

Upon observing the boy's repair job on the old flintlock, the blacksmith took Johnathan under-wing and taught him a great range of skills, paying him enough along that his father agreed to let him do his farm chores in the mornings and at night. In a short while, young Browning was competent at hand-forging and tempering, and there was essentially nothing that he could not repair (Browning and Gentry 3).

Nashville

At age nineteen Johnathan had worked with the blacksmith long enough and gained sufficient experience in metal-working and repairing weapons that he considered himself a fairly good gunsmith.

His next break occurred when someone brought in a gun made by Samuel Porter in Nashville, a good day's ride away. After studying the weapon at length, he made a bold decision and rode to Nashville to meet Mr. Porter, who was immediately impressed with the young man's experiences and mechanical acumen and hired him on in the gun shop.

After three months of working with Porter, Johnathan felt that he had learned enough fundamental gunsmithing skills that he could return to Brushy Fork, where he could carry on with his own business and where a girl he was interested in awaited.

His tenure with Porter had been most productive, though, and he came away not simply with skills he had learned in the shop—not the least of which was an ability to make gun barrels—but also with a significant accumulation of tools. By the time he had loaded his horse with his purchased tools and those Porter graciously gave him, among them boring and rifling tools for barrel making, his horse was fully loaded (9).

Illinois and Iowa

In 1824 Browning bought a small house that had behind it a shed, which became the shop out of which he would work for the next ten years, repairing and making guns. He married Elizabeth Stalcup in 1826, and the couple had their first child a year later.

After his father died in Illinois in 1833, the next

year the couple loaded their wagons with tools and household goods and their five children for the trip to Quincy, Illinois, and Johnathan's next phase as a gunsmith.

It was there in the bustling town of Quincy that Johnathan, who had established a much better shop than the one back in Brushy Fork—though equipped with only a forge, anvil, foot-powered lathe, and hand tools—invented his first percussion repeating rifle, using a very simple sliding-breech mechanism whereby a metal bar with five or six loaded chambers advanced through the receiver. After a round was fired, the bar was shifted to align the next chamber with a bottom-mounted hammer for firing.

Foot-Powered Lathe Used by Browning (Browning Museum)

The little rifle, which Johnathan never bothered to patent, brought him considerable local fame and doubtless would have been a great commercial success, had metallic cartridges not soon come onto the scene.

Though Johnathan Browning prospered in Quincy, when he fell in with Mormon Prophet Joseph Smith in 1838 and converted to Mormonism, townspeople turned their backs on him, so in 1840 he sold his gun shop and moved the family to Nauvoo, Illinois, where he joined the Mormon Community and opened a new shop.

Johnathan's interest in his new-found religion grew with increasing zeal over the next few years, and when in 1846 Brigham Young left Illinois to avoid religious persecution, he hired Browning to make guns and agricultural implements for the trip west. Johnathan loaded up his family and what possessions they could carry and headed off to Council Bluffs, Iowa, losing essentially all equity in his house and shop in Nauvoo.

The Browning family settled in the Mormon community in Council Bluffs, where he repaired firearms for settlers headed farther west until in 1852, with three wives and over twenty children, he joined the Mormon exodus and moved to Ogden, Utah, where, once again, he opened a gunsmith shop.

Ogden

No matter his great skills as a gun designer and gunsmith, when he arrived in Ogden, the needs of the people were such that Johnathan found himself tending to a great range of mechanical issues, so much so that his shop became a repair center for everything from guns to wagons to agricultural implements.

The shop was little more than a shed built with whatever materials were at hand, primarily rough-cut boards with battens to keep out the cold, and it was equipped with a rudimentary forge and well-worn tools that Johnathan had hauled all the way from Tennessee.

This modest little gunsmith shop became the schoolroom in which John Moses Browning learned the early skills that prepared him to become the greatest gun designer the world has ever known.

John Moses Browning

The Legend Is Born:

John Moses Browning

Given the skills of the father and the nature of the world John Moses Browning was born into and grew up in, it is not at all surprising that in time he too would be a gunsmith and gunmaker.

It is difficult to determine what his earliest exposure to guns was and how he might have reacted, but we know from many sources that very early on he began spending a great deal of time in his father's shop, often merely sitting on a stool and observing. Indeed, the shop was his schoolroom, where he sat and studied guns and tools, not books, and his playthings likewise were gun parts and tools.

One of the boy's favorite spots was the discard pile over in the corner of the shop, where Johnathan would toss broken, defective, or badly rusted parts that he figured he could never press into service again. Always the pragmatist, though, he knew that sometimes the very thing that is junk one day is precisely the part needed the next. Johnathan Browning rarely threw away any piece of metal.

John would daily search through the discard pile, lifting one piece of metal after another, asking his father what it was and what its function was in reference to the other parts of a gun. Johnathan would patiently answer the boy's questions, knowing, as he must have, that he was seeing a carbon copy of himself and doubtless recognizing the potential in the child.

Frequently when Johnathan determined that he needed something from the junk pile, he would describe to the boy the part he wanted and John set about finding it. If it needed to be cleaned or scrubbed free of rust or straightened, the boy would do it without being told.

Bench in Browning Workshop (Browning Museum)

In time John set up his own little workbench by the junk pile and began making things that were an improvement over what was being used. For example, he realized that one of the reasons fish were getting off lines so easily was that the hooks they had been using were made from simple household pins that were bent to shape. He reasoned that something was necessary to make it difficult for a fish to free itself of the hook, so he fashioned his own fishhooks, complete with barbs, thereby vastly improving their effectiveness.

It may be said, then, that John Moses Browning's career began right there at his little workbench by Johnathan's junk pile. He felt more at home in the shop than anywhere else, and he was quite handy, greeting customers when Johnathan was gone and writing down what they needed to be done, Though he did attend school occasionally, to his mother it appeared that he went only enough to learn how to write repair tags for the shop (32).

During these early years in the shop John learned terminology and concepts relating to guns and things mechanical in general, and he took a personal interest in items that were brought to the shop for repair. In time, when customers came to the shop, they would find themselves often discussing their gun problems with this young gunsmith, who seemed to know the parts of a gun and how they worked almost as well as the father. Indeed, often the boy would determine the problem with a non-functioning weapon before Johnathan even touched the gun.

The First Gun

The junk pile in the shop was a source of endless fascination for John, and he was forever poking through it, studying the bits and pieces he ran across or rummaging for something Johnathan asked him to find.

One of the tools he used to root around with in the discard pile was a flintlock musket barrel that had been crushed on the end. This old barrel, which he sawed the end off of, became the platform for his first weapon design.

He affixed to the barrel a piece of wood that he found lying in the shop and chopped and whittled into a shape that suited his purpose and then fashioned a flash pan beneath the vent hole where he would place his initial charge.

Lacking a hammer and trigger, the device that he ended up with still suited him enough that he was ready to give it a try. With his brother Matt along with a bucket of glowing coals and a pine stick, John carried his newly made gun into the nearby brush to try it out. He also carried enough powder and shot pilfered from Johnathan's stash to charge the weapon, which in due time he did.

When the boys were far enough away that they judged they wouldn't be seen or heard, John loaded and primed the gun, and they set about trying to find a target, which they did soon enough: three prairie chickens, two close enough to present a single target.

Holding his gun in a firing position, John aimed at the birds and nodded for Matt to light the pine branch with the bucket of coals and touch the lit end to the pan of powder, which he did, except that he bungled the effort.

On his second attempt, though, the tip of the burning stick touched the pan of powder, and it flashed, sending fire down into the main powder charge, just as John had intended. The little gun roared and threw its shot into the clump of prairie chickens, producing so much smoke and dust that at first the boys could not assess the results of their shot.

What they found when the smoke and dust cleared was one dead bird and two badly crippled ones, which Matt set off in pursuit of. John Moses Browning's first gun had been a success.

John Browning, Apprentice

Before John was even in his teens, his skill in repairing things was so widely acknowledged that it was not uncommon for him to be summoned by ladies in town to repair a pot or pan or sewing machine. When something needed fixing, John Moses Browning could probably fix it.

His skills in the shop had grown markedly over the years, and Johnathan relied on the boy to replace barrels, braze and weld, and repair broken parts. He could disassemble and reassemble any gun that he laid hands on. More and more Johnathan depended on him to handle affairs of the shop.

The turning point for the boy came one day when a freighter brought in an old percussion shotgun that a large box had fallen on and rendered useless: barrel crushed, stock mangled, almost every part of it damaged to some degree or badly worn. Upon examination, Johnathan concluded that the expenses involved in repairing the gun would surpass the cost of a new one. It was, in short, an item for the discard pile.

The freighter bought a reconditioned gun from Johnathan

and in leaving asked whether John would like to have the damaged one. The boy didn't have to think: He accepted the mangled shotgun gladly, his mind already envisioning ways to repair it.

At first angry and frustrated that anyone would allow a beautiful gun to be so terribly defiled, gradually he began to examine the shotgun very carefully, focusing his entire attention on it. Finally an idea came to him. He described the moment this way:

> A good idea starts a celebration in the mind, and every nerve in the body seems to crowd up to see the fireworks. It *was* a good idea, one of the best I've ever had, and so simple it made me ashamed of myself. Boylike, I had been trying to do the job all at once with some kind of magic. And magic never made a gun that would work. I decided to take the gun apart, piece by piece, down to the last small screw, even those that were mashed and twisted together. And when I finally did, finishing long after supper that night, the pieces all spread out before me on the bench. I examined each piece and discovered that there wasn't one that I could not make myself, if I had to. . . . If I had been in school that day, I would have missed a valuable lesson. (47)

Though the job occupied all John's spare time for the next few weeks as he meticulously worked on the little shotgun—repairing parts when he could, making new parts when he couldn't—the result was a finished product so impressive in fit, finish, and function that it might well be considered the turning point at which the boy apprentice had become gunsmith and gunmaker.

John Moses Browning, Gunsmith

The freighter's reconstructed shotgun marked a decided change in Johnathan's attitude toward the growing boy. Though it was clear that the father was still in charge and stood very tall as the gunsmith he had been for decades, it was also clear that John had earned the privilege of being treated more as an equal than just an apprentice. By the early 1870s, the former helper was running the shop.

In 1873, after a quickly extinguished fire flared up in the cluttered old shop, John took over reconditioning the place, making it a more suitable working environment for all concerned. At this point he was essentially in full charge of things, with Johnathan popping from time to time as the spirit moved him or when John needed him for consultation.

Work went on as it had for over twenty years in the renovated shop, with gun repairs remaining the bread-and-butter for the Brownings, but John had been mulling over for years the notion of building a new gun, one of his own design and of his manufacture. He had learned a great deal about guns from repairing them, but often he would find himself thinking that he could improve on some of the parts he handled. He also found himself lamenting the fact that Johnathan had lost his enthusiasm for doing much more than working on other people's guns once he got to Utah.

One day in 1878, while John was working on an old single-shot rifle, the parts spread out on the bench before him, he called Johnathan over from his favorite perch in the shop, the anvil, and pointed to the array. He declared that he could build a better gun.

"I know you could, John Mose. And I wish you'd get at it. I'd like to live to see you do it." (59)

That little bit of encouragement was all it took. One evening John sat at the kitchen table and sketched out the design of his first real gun between bites of butter and biscuit, and in a year he had built the gun and applied for a patent. He was twenty-three.

The First J.M. Browning Rifle

Only someone knowledgeable in metallurgy and experienced in metalworking can appreciate the challenges that John faced in building that first model. He had to hammer-forge most of the parts, honing them to fit with chisel and file, using the shop's old breast drill and foot lathe to drill and thread. In building the receiver, he had to "honeycomb" the interior by drilling and then chip out the webbing with a chisel.

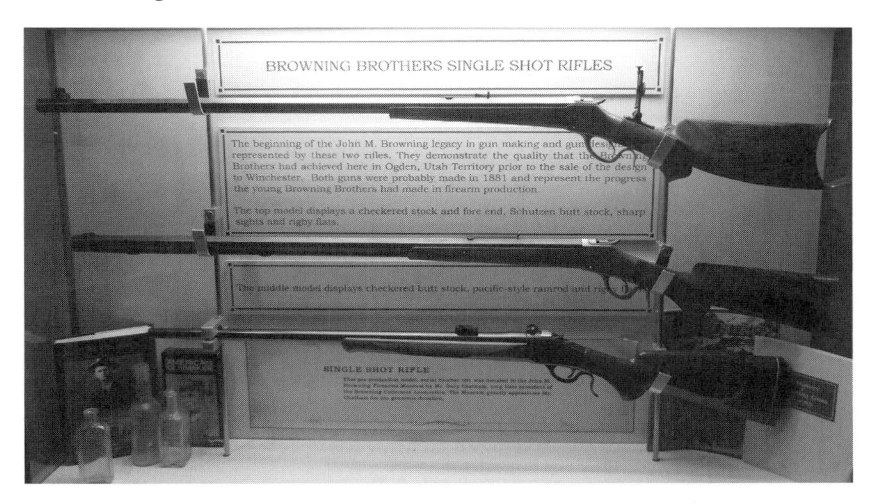

Browning Single-Shot Rifles (Browning Museum)

Arduous as the task must have been, within a year after the birth of the idea for his first gun, he had finished the model and applied for a patent, which was issued in 1879. The barrel marking reads, *JM Browning Ogden U.T. 1878.*

Browning Store Front (Browning Museum)

The Browning Brothers Hang Their Shingle

The Factory

John knew that as fine as his new single-shot rifle was, if he intended to make anything of it, he had to have a means of producing it, and the confines of the new shop and its limited machinery made that essentially impossible. He entertained the notion of selling the patent to some firm back east for production, but he was sufficiently independent of mind and certain of his capabilities that he decided to go ahead and build the rifle himself. This meant that he had to have a larger and better-equipped shop in which to make the new gun, and that became his next project.

He bought a little piece of property only three blocks from home, and the Browning brothers—John, Ed, George, Matt, and Sam—set about building the new factory, using what little carpentry and masonry skills they had to fashion a twenty-by-fifty-foot building, large enough to accommodate the equipment John figured he would need to start production and allow room for a sporting-goods section,

As time permitted, while they worked on the new factory and

kept the old shop open for business, John went over with his brothers the structure of the new rifle, teaching them every little detail of each piece.

By the time the building was finished and equipped and functioning, John had already taken four orders for his new rifle, coming solely from people who had handled the model. He had taken on another gunsmith, Frank Rushton, who happened by one day and decided to stay, and the two of them, with whatever assistance the less experienced brothers could lend, began setting up shop.

John arranged to have outside manufacturers produce barrels and forgings for the new rifle, and he divided the labor according to the capabilities of the brothers, with Ed doing the milling and machine work, Matt working on the stocks, and Sam and George tapping centers and drilling holes. Rushton and John did much of the final fitting.

Initially John had stockpiled sufficient materials to produce a hundred rifles, and within three months the factory had finished twenty-five copies, all meticulously inspected and function-tested by John himself, who from time to time would delicately apply what a friend of his referred to as a "feather touch" with a file. Everything had to be perfect in the master's eye. (86)

A week after the rifles passed his careful inspection, all of John's rifles were sold, and he had orders for several more. He also had enough cash on hand to go ahead and lay in a supply of sporting goods, which, like the rifles, sold out almost immediately. Browning Brothers was a success.

The First Repeater

In spite of the success of his new single-shot rifle, John Moses Browning could not be content: Within two years of the opening of his factory, he had designed a repeating rifle with a tubular magazine, which he acquired a patent for in March of 1882. In September of 1882 he applied for and received a patent for a second repeating rifle.

What he had in mind—and what he continued to work on

as time permitted—was a repeating rifle, an ingenious design that would become the revolutionary Winchester 1886.

Winchester had developed and produced a line of repeating rifles that all but wrapped up that market, but their guns were designed for lighter calibers, such as the .45 Long Colt and .44-40. John's repeater, like his newest single-shot, was so robustly built that it could handle essentially any cartridge in existence.

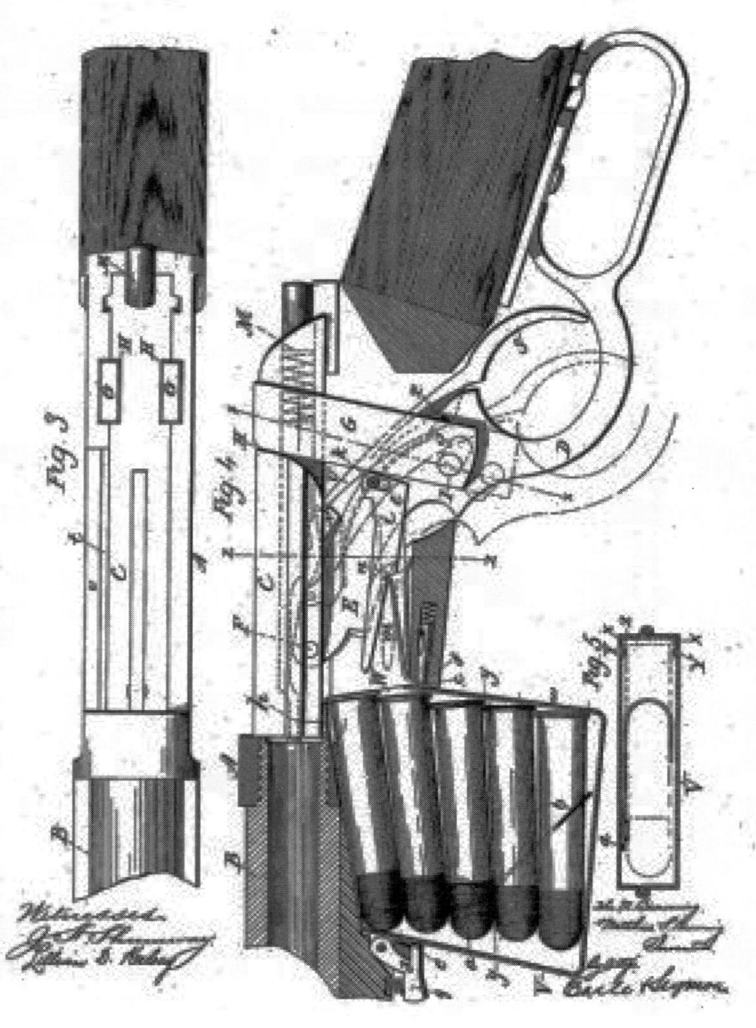

Winchester 1892 Patent Drawing

The Winchester Years

Enter T.G. Bennett

In 1883 a Winchester salesman happened upon a well-worn Browning Single Shot in .45/70 caliber and was so impressed with it that he bought the rifle and sent it to the Winchester factory to see whether they might be interested in it. Upon examining and testing the weapon, they decided that they were very much interested in it.

Vice President and General Manager of Winchester, T.G. Bennett was sufficiently impressed with the rifle that he made the trip to Utah and entered into negotiations with John, out of which came an agreement: Winchester would pay John $8009 for production rights to the gun and put Browning Brothers on the Winchester jobbing list, which would require expansion of the factory but pay off in significant dividends (100-01).

Winchester brought the falling-block rifle out as the Winchester 1885, and it enjoyed immediate and immense success. It could handle any load a shooter could shove into it, from the .22 Short to the .50/95 Winchester Express, and Winchester produced the gun in a broad range of calibers until 1920, when sales suggested that the repeater was preferred by most buyers.

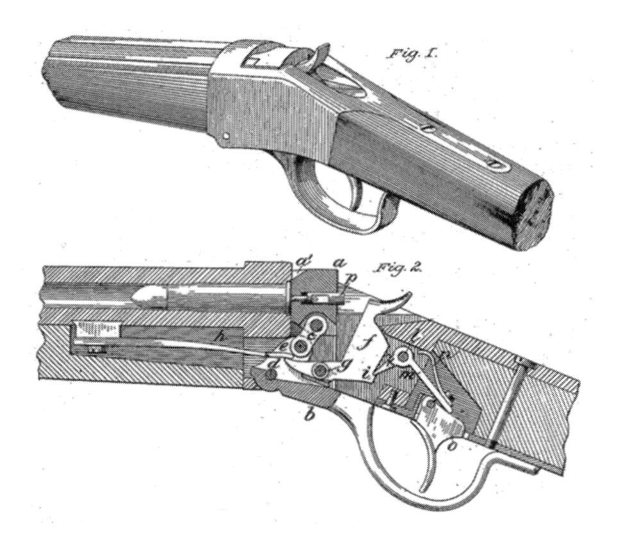

Browning Single Shot (Winchester 1885)

Though the deal with Bennett allowed Winchester to manufacture and market the Browning Single Shot, John saw no reason why he couldn't continue to build and sell the rifle himself. He had a lot of stock parts on hand, so the little Browning factory kept churning out the gun until word got back to Bennett, who sent a stern letter suggesting that the Brownings were in violation of their agreement and should cease production immediately, which they did (103).

The Winchester 1886 and Successors

All the while the Browning brothers had been forging ahead with Single Shot production, John had been turning over in his head that revolutionary design for a repeater that would eventually become the Winchester 1866.

Winchester had cornered the market on repeaters with their lever-actions, but their guns were not designed to accommodate the heavier loads being introduced year after year. Since John's Single Shot action could handle any cartridge being made at that time, he was determined to design a repeater that could equal that performance.

In May of 1884, John applied for a patent on the new lever-action rifle, which was initially chambered for the .45-70 round but would later be chambered in such powerful cartridges as the .50-110 Express, favored by buffalo hunters. He wrapped the model in simple brown paper, and he and Matt carried it with them to deliver it in person to Bennett, who is speculated to have bought the patent for $50,000, a sum that John said was "more money than there is in Ogden" (108).

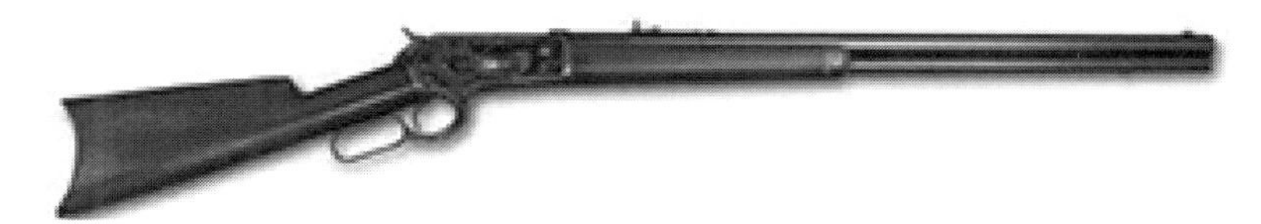

Winchester 1886

Like the Single Shot, the 1886 was an immediate commercial success for Winchester, further strengthening their monopoly on the repeater market. When smokeless powder was introduced, the 1886 action was strong enough to accept the powerful new cartridges, requiring only a stronger barrel to handle the heavier loads. (Winchester introduced this strengthened version of the 1886 as the Model 71 in 1935 in the powerful .348 Winchester round, and it was manufactured until 1958.)

When a bit later Winchester came to John and asked whether he could design a smaller-scale lever-action to replace the 1873, he designed the 1892, which accepted common pistol cartridges of the day: .32-20, .38-40, and .44-40. His next design was the enormously successful Model 1894, with over a million and a half sold before 1914 and a favorite of American deer hunters to this day. Still much smaller than the 1886, the Model 1894 was designed to fire powerful smokeless-powder cartridges, and though it has been produced in many different calibers, the one that it is most strongly associated with is the .30-30 round.

Winchester 1894

Winchester 1895

Winchester's enthusiasm for John's new designs was insatiable. His next project, the Winchester 1895, was a box-magazine lever-action strong enough to handle the heaviest cartridges in existence, up to the very powerful .405. (135) A favorite of Theodore Roosevelt, the rifle became an immediate success with big-game hunters all over the world, and it saw military use in the Spanish-American War.

The Winchester 1890

Over the years Winchester had produced their popular Model 1873 in many different calibers, including the .22, but for some reason—perhaps because of its weight—the 1873 .22 simply did not appeal to buyers, so Bennett got in touch with John and asked whether he could develop a .22 rifle for Winchester.

John immediately started to work on the project and in 1888 he and Matt had a patent issued to them for a pump-action .22 rifle. Though Bennett was a bit leery of the pump-action of the rifle, he recognized the genius of the mechanical design and accepted the fact that a shooter could operate the pump action more quickly and with better accuracy than he could a lever-action gun. So it was that Winchester bought the patent, and the 1890 was born.

Winchester 1890 .22 Gallery Gun

In time the 1890 would become extraordinarily popular to the American public, and it became so ubiquitous in shooting galleries that eventually it was often referred to as the "gallery gun."

The Model 1890 initially sold for $5.00, and it went through nine reconfigurations over the years and sold almost a million and a half copies.

A Better Shotgun: The 1897

When T.G. Bennett came to John about designing a new shotgun for Winchester, what he had in mind was a lever-action built along the lines of the '86, but what John brought to him after a few months was something more akin in design to the 1890 .22. It was a model of the twelve-gauge exposed-hammer, pump-action model 1893.

Whatever Bennett's initial misgivings about a pump-action shotgun, he certainly appreciated the success of the 1890, and he was impressed enough with the new design to buy it and put it into production. Winchester manufactured the 1893 until John introduced the Model 1897, an improved model with a thickened receiver, making it suitable to fire shells charged with smokeless powder.

The super-reliable 1897 quickly became one of the most popular shotguns of all time, saturating the civilian market and going on to serve as a very effective combat weapon in World War I and as a favorite weapon of law enforcement and prison guards.

Winchester 1897 Trench Gun

The Break with Winchester

The last gun that John designed for Winchester was another single-shot .22, this one a bolt-action boy's gun, which could be manufactured and sold for a very low price.

In 1898 John made a radical departure from his usual designs and developed a semi-automatic shotgun, the Auto 5, for which he received a patent in 1900. By this time, he had been dealing with

Winchester for nineteen years, selling them forty-four guns, with ten models actually produced and sold, so he naturally approached T.G. Bennett with his new shotgun.

By this point in his history with Winchester, John felt a little more confident in his dealings with Bennett, so in his negotiations over this new gun, he asked for a financial arrangement he had never asked for before and something Winchester had a set policy against: royalties. Accustomed to lever- and pump-action designs, Bennett was very much uncertain of the potential in the Auto 5 and declined John's proposal. Doubtless it was a decision he would regret for the rest of his life.

Remington expressed an interest in the Auto Five, but the very day John was supposed to meet with Marcellus Hartley, president of Remington, to seal the deal, the man died suddenly of a heart attack, throwing everything into a state of confusion. He ended up taking his patent to Fabrique Nationale, in Belgium, where he found the gun warmly welcomed. They struck a deal, and the Browning Auto Five was introduced to the world. It is still being manufactured today.

Thus John Moses Browning's highly successful nineteen-year-long association with Winchester came to an end.

The Browning Machine Guns

Harnessing the Energy

The idea for a fully automatic weapon came to John one day in 1899 while he and his brothers and some friends, all members of the Ogden Rifle Club, were at the range they used near the river.

In the weeds of the overgrown range, John noticed something that doubtless he had seen before but never really taken careful note of: Well out in front of the person who was firing at the time, John noticed how the weeds were stirred by the blast from the rifle.

John was through shooting for the day. What was on his mind as he headed back to the shop was a notion on how one might harness that wasted energy from a weapon being fired and utilize it to operate the action.

Back at the shop, he drilled a hole slightly larger than the diameter of a .44 bullet in a piece of 2x4 and set the board on the floor. He rigged up a Winchester '73 just a quarter inch from the board, with the muzzle lined up with the hole he had drilled. He was convinced that the board would do what it did: When he pulled the trigger with a length of wire, the piece of 2x4 caught the force of the blast and skittered across the shop. His experiment had confirmed his notion that when a gun is fired, it produces a great deal of wasted energy that might well be harnessed and put to use. (144-45)

His next step was to fashion a device on the '73 that he had initially used in his experiment, running an operating rod from the modified lever to a disk with a hole slightly larger than diameter of the .44 round. The disk was mounted on the muzzle so that the bullet would pass through the hole, and the excess energy generated by the round being fired would force the disk down, actuating the rod attached to the lever. It worked: When he fired the rifle, the action functioned as if a human hand had operated it.

The '73 setup was quite cumbersome, so John abandoned it and built a series of automatic weapons in the months to come, improving on his newly discovered principle of a gas-operated action. A year and a half after the first experiment, he filed patents for two machine guns using slightly different principles of operation, and a year later (1892) he filed a patent for a gun in which a hole was drilled in the barrel to take advantage of the propellant gases, directing the energy against a piston. The piston pushed a rod that operated the action. Over the next three years he filed four more patents on this weapon.

Always eager to devise and introduce new designs, John was working away at any number of projects. In addition to continuing to refine his machine gun projects, he began working on a series of handgun principles that would in time become concepts that altered forever the way semi-automatic handguns were built. In October of 1896 he filed three new patents relating to handgun operating systems: the blowback, the "locked-recoil system with a turning lock," and the "locked-recoil system with a pivoting lock" (148).

Passing Muster with Colt and the Navy: The 1895

Just over a year after his experiment with the Winchester '73 flapper-and-rod apparatus, John approached Colt regarding his latest machine gun design and asked whether they might be inclined to observe a demonstration of his new weapon. Since Colt had some experience in the field of machine guns, notably with the Gatling Gun, they seemed a reasonable choice for production of his gun.

Prior to this, Colt had had no contact with the Browning brothers, but they certainly knew about the Brownings and the fine weapons they had been building, so the president of the company, John Hall, extended them an invitation to visit with him.

As ragged as this first model he carried to Colt might have been—heat-blackened, with hammer dents and machine marks all over it—John and Matt lugged the ugly duckling into the office of Hall, who, after a period of conversation, graciously consented to allowing the Browning brothers to demonstrate their new machine gun.

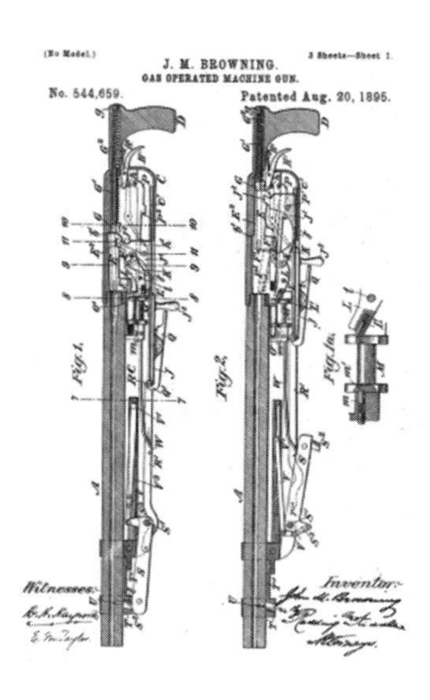

Patent drawing of Browning 1895

They had brought four belts of ammo with them, a total of 200 rounds of .45-70, which their little gun ran through, without so much as a hiccup, in a matter of seconds. Though at the time he did not commit to anything, Hall asked John to leave the machine gun with him so that he could show it to some of his military contacts and perhaps demonstrate it before them.

John decided not to leave the gun with Hall, but he did promise to return and conduct a demonstration himself before the officers. He fully intended to take the gun back to Ogden and spruce it up a bit, but Hall argued against any cosmetic changes: “When you set up the gun in the firing room, I nearly laughed out loud, but at the end of two hundred rounds, I felt like giving three cheers. If we get some officers in, put on the same kind of show for them. It's a great show" (154).

Hall did indeed interest some of his Navy contacts in the gun, but true to military form, the Navy insisted on having things done on their terms: The gun would have to fire 1800 rounds in three minutes (600 rounds a minute) without pause.

John with the 1895 (Browning Museum)

So it was that the Browning brothers arrived in Hartford bearing the machine gun and 1800 rounds of .45-70, strung in belts made by an Ogden tent maker and layered in nine wooden feedboxes. The audience: Hall, one of his managers, and two Navy officers.

Perched on a bicycle seat he had mounted to one of the legs of the tripod, with his boxes of ammo within easy reach, John unleashed his ugly little machine gun, which fired 1800 rounds without a single failure to fire, much to the delight of those on hand (158).

John did not sign a contract after the successful testing of his machine gun, though Hall offered one: He preferred to continue to improve on the design until he had a model that better pleased him. Within a year he presented Hall with his new model, which was put into production in 1896. Chambered in the .30/40 Krag and 6 mm. Lee cartridges, the Colt 1895 Automatic Machine Gun was tested in 1896 and approved by the Navy, which ordered fifty copies of the gun and put them into operation in 1897.

The 1895, nicknamed The Peacemaker after its performance during the Spanish-American War, saw action during the Boxer Rebellion, and Colt sold copies of the weapon to South American and European countries. It was also called the Potato Digger, a reference to its downward thrusting recoil mechanism that would actually gouge into the dirt if fired at too low a level.

The 1895 was only the beginning of a long string of successful Browning machine guns, with ultimately more than sixty of his "automatic machine guns" adapted for use on land and sea and in the air (172).

The 1917 BMG

In June of 1900 John filed a patent application for a machine gun initially referred to as the Browning Model 1901, the year the patent was awarded. After securing the patent, he did not work on it again until 1911, when he modified it to use a water jacket, permitting the gun to be fired for longer periods of time without overheating. Similar in design to the popular Maxim, though more than fifty pounds lighter, the new Browning machine gun used the .30-06 cartridge, the same caliber our Springfield and Enfield rifles used.

After the declaration of war in 1917, the U.S. military realized that their existing arsenal of machine guns was limited in number and essentially obsolete, so they set about arranging a series of tests to find a weapon that would better equip our troops to engage in modern warfare. Easily besting its competitors in the grueling tests that followed, the Browning 1917 entry fired a total of 40,000 rounds without a mechanical failure. The Ordnance Board was still not fully convinced, so a second 1917 ran through 21,000 rounds, firing uninterrupted for just over forty-eight minutes (165).

Orders for the 1917 came almost immediately after the cessation of trials, but because of the late entry of the country into World War I, the Model 1917 was not used by American troops in combat in France until September, 1918, and saw only limited service during the war.

Responding to requests from General John Pershing, Commander of the American Expeditionary Forces in Europe for a larger-caliber machine gun, Browning set to work altering the 1917 to fire the much larger .50-caliber round, which was essentially a vastly enlarged configuration of the .30-06 round. In September of 1918 the first model of the .50, built by Colt, was successfully tested.

John with the 1917 (Browning Museum)

Subsequent modifications of the 1917 resulted in improved machine guns that saw service in World War II, the Korean War, and the early stages of the Vietnamese War.

The 1919 BMG

All the while, in typical John Browning fashion, John was working away at improving his machine guns. Realizing that the weight and cumbersome nature of a water-cooled automatic weapon limited its use in all the services, he set about developing an air-cooled .30-caliber machine gun, which utilized the same tried and true functioning principles of the 1917 but weighed much less and was cheaper to produce.

Firing the .30-06 initially from a fabric belt, the 1919 came in at just over thirty pounds and fired 400-600 rounds per minute, usually from a tripod mount.

In time this weapon, manufactured in myriad configurations and calibers, would become ubiquitous in the military theaters of the world, serving on land and sea and in the air. It was a primary medium machine gun in all our armed forces throughout World War II, the Korean Conflict, and the Vietnamese War and is still in use among combat units across the globe today.

Model of 1919 (Browning Museum)

Though the 1919 weighed just over thirty pounds, in typical infantry application it required a mount that weighed just as much, making it at least a two-man machine gun. A third man was charged with carrying ammunition to keep the weapon running. At a fire rate of up to 600 round per minute, the barrel overheated fairly quickly, so the firer had to pace himself when using it in full-auto mode.

The Ma Deuce

After the U.S. entered World War I and our forces began to experience the realities of mechanized warfare, AEF Commander General John Pershing realized that our rifle-caliber machine guns were not sufficiently heavy to penetrate armor of any significant thickness, so he turned to the Army Ordnance Department to develop a weapon to address the problem.

The AOD's response was to convert some test Colt machine guns, using the French 11 mm. round, but that round lacked the muzzle velocity Pershing sought, 2700 fps., and it was a rimmed round, which General Pershing did not want.

Since 1917 John Browning had been experimenting with a larger machine gun round, in configuration simply a much larger version of the .30-06 cartridge. When he was approached about designing a larger-caliber weapon, after quizzing government officials at length about the nature of the round they wanted, he set about creating a weapon to fire this new cartridge. As absurd as it might sound, John had the machine gun built before the ammunition to test it was delivered.

In September of 1918 John personally conducted successful field tests of his new .50-caliber, which was essentially a beefed-up version of the water-cooled 1917. The weapon performed admirably.

Since he had already been experimenting with a much lighter air-cooled version of the 1917, John decided to make similar modifications to the .50, and the legendary M2 was born.

Weighing over eighty pounds, plus tripod and ammunition, the infantry version of the M2 definitely fell under the category of a heavy machine gun, requiring a crew of four or five men to be transported, set up, and operated effectively, and it has served in its ground-based capacity in most branches of our military since 1933.

The M2's variants are legion, serving on land and sea and in the air in almost every capacity imaginable. Its role as an aircraft armament sets it apart as one of the most ubiquitous aerial weapons ever employed, and a number of allied aircraft were fitted with it, from fighters to heavy bombers.

Browning M2 Machine Gun (Browning Museum)

Operating at rates of fire up to 1200 rounds-per-minute, depending on its configuration, the M2 fired an impressive array of projectiles—armor-piercing, incendiary, armor-piercing incendiary, and tracers—making it extremely versatile in battle.

Almost a hundred years from its introduction, the Browning M2 is still a widely used heavy machine gun today.

II

THE BAR:

A LONG, DISTINGUISHED HISTORY

John and Frank Burton, of Winchester, Examine BAR (Browning Museum)

The BAR is Born:

Out of the Forge of War

The Heavy Machine Guns of the War

When the United States entered World War I in April of 1917, our expeditionary forces were ill-equipped to become engaged in modern warfare. Our total inventory of machine guns amounted to just over 1300 weapons, most of them Maxims, Lewises, and French-made machine rifles, none of which were regarded as suitable for going up against well-armed enemy troops entrenched in European battlefields.

Repudiating existing notions that a machine gun could function effectively only through some mechanical means of operation—a hand crank, for instance, in the case of the Gatling—Hiram Maxim observed, as did John Moses Browning, that the energy expelled by ignited cartridges might well generate sufficient force to operate the mechanism of the Maxim.

Water-cooled and fed by a belt of .303 British ammunition, the Maxim fired up to 600 rounds per minute, giving it the equivalent firepower of between eighty and a hundred rifles of the same caliber. The muzzle velocity fell well short of the requirements of American military authorities, rendering it at best a medium-weight machine gun. Due to its construction, with a water-jacket and reservoir, the weapon was in fact quite heavy and required a minimum crew of four to maneuver and operate effectively on the battlefield.

Though the British expressed an immediate disenchantment with the Maxim, the Germans readily embraced it, so that when World War I began, the British and French had in their arsenals only a few hundred Maxims, as opposed to well over ten thousand in the hands of German troops, a number that would surpass a hundred thousand before the war was over.

The British developed their Vickers machine gun from the Maxim design and likewise chambered it for belt-fed rimmed .303, adopting it as their primary automatic weapon in 1912. Though still a heavy weapon requiring a crew of three men for successful battlefield applications, the Vickers nonetheless weighed less than the Maxim.

The Lewis machine gun was introduced as a lighter automatic weapon in 1915, and the British pulled their Vickers from infantry units and redesigned them as heavier weapons firing a 12.7 mm (.5") round. Instead of requiring a belt of ammunition for operation, the Lewis was fed from a "pan" magazine, which held either 47 or 97 rounds of .303 and sat atop the rear of the receiver and could easily be managed on the battlefield by two men.

When American troops entered combat in World War II, they were issued British or French machine guns or provided the 1917, which was generally not favored over the Maxim or Vickers.

The Need for Walking Fire

As effective as the medium and heavy machine guns might have been in holding defensive positions or providing suppressive fire, they were not easily maneuvered in the heat of battle and might require up to six men to situate them and keep them firing. Unless mounted on some sort of vehicle, they were quite ineffective in attacking enemy positions.

Trench warfare, which characterized a good deal of the combat of World War I, required assaults across contested territory with essentially nothing but rifles, handguns, and shotguns, giving attacking troops very little suppressive firepower as they attacked enemy positions. What was needed was a lighter weapon giving a single soldier the firepower of a machine gun as his unit made an assault on an enemy position. The term used for this kind of offensive warfare is "marching fire" or "walking fire," and it requires a heavy volume of sustained fire to keep the enemy pinned down until the final charge.

Existing machine guns were simply not mobile enough to utilize during such offensive maneuvers. The two lightest machine guns available were the Lewis, which weighed over thirty pounds, and the French Chauchat. Because of its lighter weight and ready availability, the Chauchat became the light machine gun of choice, and the U.S. purchased 16,000 of the guns from the French and issued them to our infantrymen.

French Chauchat

Weighing in at around twenty pounds, the Chauchat fired the 8 mm. Lebel rimmed cartridge from an open-sided curved magazine that was notorious for failing to feed properly. When lubricated to facilitate movement of the rounds through the magazine, mud and sand easily entered the openings along the sides and jammed it. In addition to a history of failure of magazines and internal parts, with its long-recoil system of operation, the Chauchat had an extremely slow rate of fire: 250 rounds per minute.

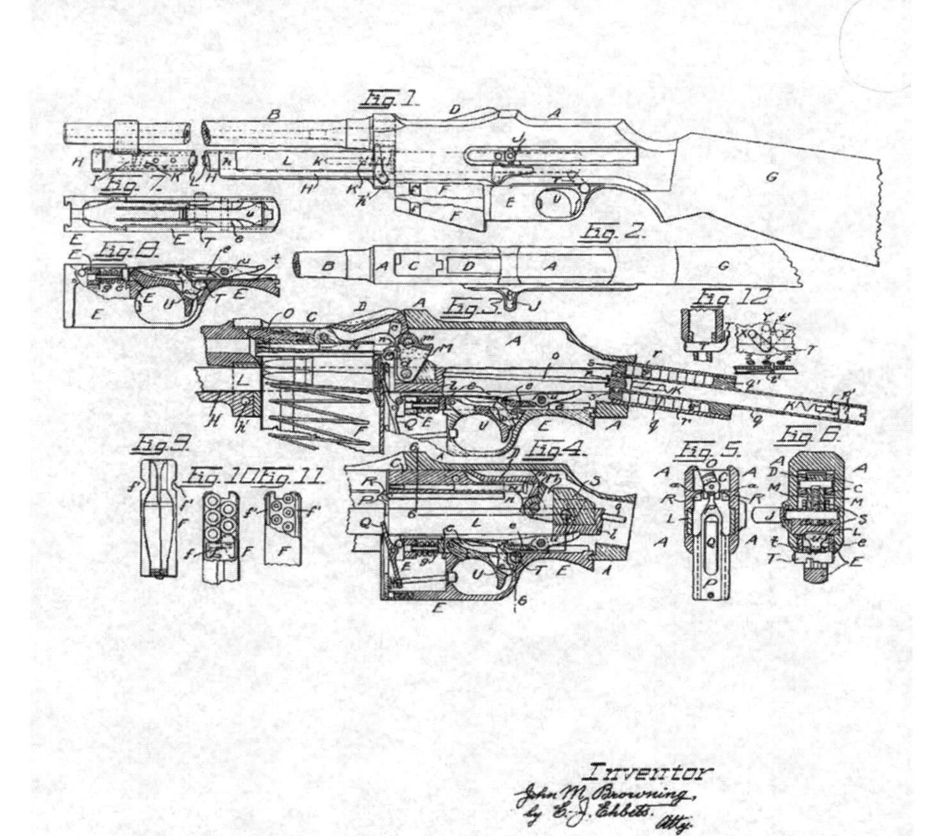

Patent Drawing of BAR (Browning Museum)

In short, the American doughboys were ill prepared to engage in the kind of warfare expected of them, and they uniformly despised the French Chauchat.

John Moses Browning to the Rescue

Convinced that sooner or later the country would be drawn into the war in Europe, John began concentrating on improving a weapon he had been at work on since as early as 1910: an automatic rifle weighing less than twenty pounds and firing the .30-06 cartridge. This new machine gun, equipped with a buttstock and forearm, was gas-operated and air-cooled and had a firing rate of over 500 rounds per minute.

In order to make it truly mobile, this gun was designed to utilize twenty-round magazines instead of the bulky belted ammunition of most other machine guns of the day. Unlike the troublesome magazine of the French Chauchat, the Browning magazine was a steel box with spring and follower and open only at the top, where it was inserted into the gun. Though limited in round capacity, the loaded magazines could be carried in large numbers in bandoliers, which could be worn around the waist or draped over the shoulder. A soldier with two bandoliers of magazines, plus a charged one in the gun, could carry 500 rounds of ammunition.

Furthermore, should a BAR gunner run out of ammunition, he could simply borrow rounds from soldiers bearing rifles of the same caliber and load them directly into his magazines instead of having to try to string together belts of ammo to keep his weapon running.

The twenty-round magazine also taught the BAR man to fire in short bursts and conserve his ammunition, at the same time improving accuracy and diminishing the chance that he would overheat the barrel with long stretches of uninterrupted firing.

In short, the disadvantages of the twenty-round magazine were offset by distinct *advantages*. Whether Browning thought all this through during his planning and building process, the long, distinguished history of the BAR in battle certainly proved the genius of the design.

As usual, John built the weapon with functionality and endurance in mind. The BAR was robust, with an extremely strong receiver, heavy barrel, and a quite small number of internal parts, they too designed to endure any sort of merciless combat conditions to which they might ultimately be subjected.

When the need arose, once again John was there with the solution: If our military need a light-weight machine gun to carry on the European War, he had the gun.

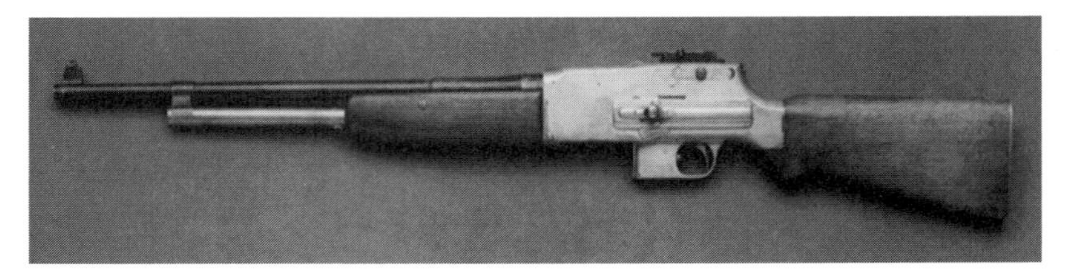

BAR Prototype (Browning Museum)

When in May of 1917 Secretary of War Newton Baker assembled an ordnance team to examine existing light machine guns and prototypes, it was not surprising that the Browning competition entry won immediate acceptance and was adopted as the Browning Automatic Rifle, fondly known simply as the *BAR* from that point on.

The 1918 BAR Goes into Production

Though the logical manufacturer of the BAR would have been Colt, who initially shared design rights with John, the Colt plant was already running at capacity. Colt suggested building a new factory to produce the BAR, but the military argued that there was no time for erecting a new factory and training staff. The solution the government came up with was to acquire production rights to the gun from Colt and Browning and contract with Marlin-Rockwell and Winchester to build the guns in a timely manner. The contracts were issued in September of 1917.

Since no production drawings of the BAR were available, Winchester's engineering department borrowed John's working model from Colt and over a single weekend developed the blueprints necessary for them to put the gun into production. Winchester then assisted Marlin-Rockwell in tooling up to manufacture the BAR.

By December of 1917, Winchester had made its first delivery of the new machine gun, and Marlin-Rockwell and Colt started shipping in January and February of 1918. As a side note: Though the BAR was initially manufactured in 1917, in order to avoid any confusion with references to the 1917 water-cooled Browning, it was officially labeled the Browning Automatic Rifle, Model of 1918. These initial models were selective fire, permitting a soldier to fire it in either automatic (around 600 rounds per minute) or semi-automatic mode.

By the time of the Armistice, over 50,000 1918 BARs had been delivered to our troops in Europe, and its manufacture continued until 1919, with a total production of more than 100,000.

The BAR Goes to War

The BAR first saw action in mid-September, 1918, when it was placed in the hands of troops of the 79th Infantry Division during the Meuse Argonne Offensive, and it immediately proved its worth. In spite of the limited magazine capacity and a weight that made it difficult to be fired from the shoulder, Browning's contribution to the war effort was greeted with great enthusiasm by the troops fortunate enough to be issued one.

The primary attribute of the BAR that set it apart from other light machine guns of the day was its heavy-duty construction. In typical Browning fashion, John over-engineered the gun so that it would perform under the most adverse conditions imaginable in terms of physical environment and weather, and it would be hard to imagine conditions more detrimental to the proper operation of a machine gun than those experienced by troops during the trench warfare of World War I.

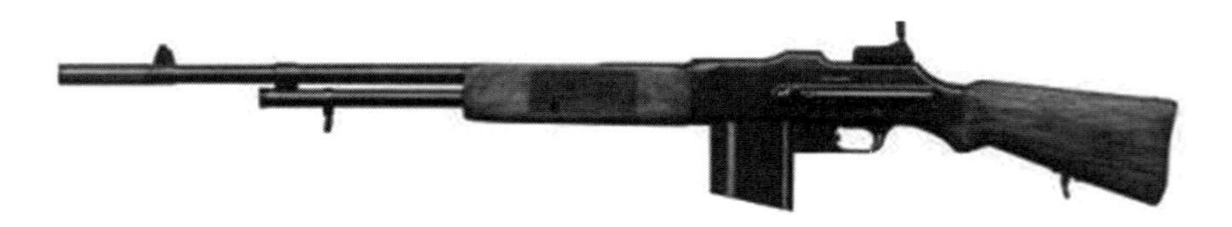

Typical Configuration of BAR during World War I

The French light machine gun it replaced, the Chauchat, was prone to mechanical breakdown from hasty production and flimsy metal stampings and the use of an open-sided magazine easily fouled by sand and mud. The BAR weighed no more than the Chauchat, but it was robustly engineered, and almost all its metal parts were machined and inherently unlikely to break or wear out. The box magazine was likewise well constructed and reliable.

Though he would never offer confirmation, it was said that Val Browning, John's son, was the first soldier to fire the BAR during the War. He certainly did use the weapon in combat, and he was always prepared to explain to others the components of the 1918 and teach them how to operate and maintain it.

Whereas it is true that in automatic mode, a magazine could be exhausted in a couple of seconds, recharging with a loaded one took only five seconds or so, much faster than the Lewis or Chauchat could be loaded, and generally the firer had no fewer than a dozen charged magazines at his disposal. Usually he would also have a companion carrying another one or two bandoliers of twelve charged magazines.

The weapon could be fired from the shoulder or slung and fired from the hip, and in later applications, it was fired prone, using a bipod, which in time became a standard accessory. Though the 1918 could indeed be fired in semi-automatic mode, typically the bearer would use two- and three-round bursts and, when the occasion called for it, fully automatic fire.

In an effort to assist the bearer in achieving more successful walking fire, an ammunition belt was designed for the BAR that featured not only pouches for magazines but a metal cup in which to stabilize the butt of the weapon while advancing on foot across the battlefield.

But for some early glitches in some BARs delivered from Marlin-Rockwell, all soon enough resolved, the 1918 performed as expected, and those who fired it had nothing but praise for it.

Because of its late arrival on the front, the BAR did not have much time on the European battlefields to prove its true merit, but its reputation for performance and durability was quickly established, and that reputation would do nothing but grow and glow in the decades to follow.

Clyde Barrow with BARs

Between the Wars:

The BAR on the Domestic Front

Experimentation on a Solid Foundation

After the war in Europe was over and the world returned to a condition of relative peace, our troops and weapons came home, the men to resume their civilian lives, and the weapons to be distributed to various training facilities across the country or to units of the National Guard or in other ways disposed of.

The BAR was, of course, among them. A number of the weapons remained in the hands of our allies after the war was over, since the weapon's superiority had by that time proven itself sufficiently well, but the majority of them were shipped back to the United States.

Though the BAR saw only limited action in the hands of Marines involved in small engagements around the world over the next couple of decades, its purpose had been served and its reputation established.

No matter how sound and successful the initial model of a weapon might be, it will in time undergo alterations as experience at the hands of those who use it dictate. No inventor can anticipate future adaptations that his brainchild will undergo. So it has always been, and so it will ever be. And so it was with John Browning's 1918 BAR.

Given the popularity of the BAR and of the horse cavalry as a viable fighting element in early modern warfare, it is not surprising that the Ordnance Department would make an effort to adapt the weapon for use by mounted troops. To improve durability during sustained fire, the newly designated M1922 BAR had a heavier barrel equipped with cooling fins, and it came fitted with a bipod and equipped with a stock rest to improve stability. The rear sling swivel was relocated to allow the mounted soldier to more easily control the machine gun. In its final configuration, the 1922 topped twenty pounds.

Since infantrymen could not possibly be expected to keep up with mounted BAR operators, special hangers were designed so that pack horses could carry extra ammunition and parts in a support role.

There is little to no evidence that the Model 1922 ever proved its merits as a combat weapon, but it remained in service for nearly twenty years.

But for the short-lived M1924 and M1925 (R75) BARs, which featured a pistol grip, a larger forearm, and a cover for the ejection port, little other significant experimentation occurred with the BAR until Colt introduced the Monitor Automatic Machine Gun in 1931.

To meet the needs of law enforcement agencies for heavier arms to combat the growing firepower of crime syndicates during Prohibition, Colt designed a lighter version of the BAR, the R80 Colt Automatic Machine Rifle, which was simply referred to as the *Monitor*. While a favorite weapon of the Law, the Thompson .45 did not have the power to penetrate the thick steel shells of the automobiles of the day, whereas the powerful .30-06 round could easily do the job.

Minus a bipod, the Monitor had a pistol grip, and Colt added a shortened forearm and lighter, shorter, finned barrel tipped with

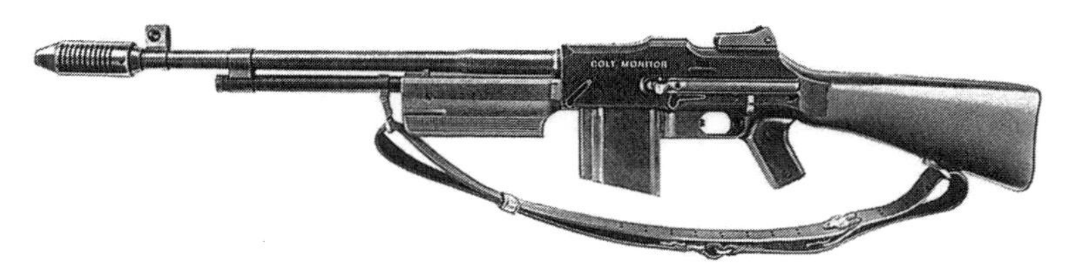

The R80 Colt Monitor

a Cutts Compensator, which reduced recoil and the gun's tendency to rise when fired in automatic mode. The weapon weighed in at thirteen pounds.

Though Colt produced only 125 Monitors, the weapon was quite popular among those who used them: law-enforcement agencies—especially the FBI—and bank and prison guards.

In 1932, a version of the 1918 designed for "bush warfare" was developed by Marine Major H.L. Smith and subsequently evaluated by the Ordnance Department. The barrel was shortened nine inches, with the subsequent relocation of the gas port and tube, and though the weapon measured just under 35 inches, the weight reduction was minimal. This new prototype equaled the 1918 in many respects, but the model was abandoned after the initial evaluation.

Among the more radical reconfigurations of the BAR between the wars was the Marriner Browning T10, a belt-fed version developed in 1933 by the son of John's brother Matthew, Marriner. Colt built the prototype of this heavy-barrel, tripod-mounted machine gun and tested it in 1834

The weapon went through several stages of evolution, with Springfield, Auto Ordnance, and High Standard involved in its development. Finally it went back to Springfield, where the last model, the T23E2, was rejected by the Infantry Board in 1943 as being not worth the further investment of time, energy, and money.

The M1918A1, introduced in 1937 in an attempt to improve the combat efficiency of the 1918, was an experimental version of the BAR equipped with a cut-down forearm, hinged butt plate, and hinged, spiked bipod mounted on the gas tube. It retained the semi-auto and full-auto capabilities of the 1918.

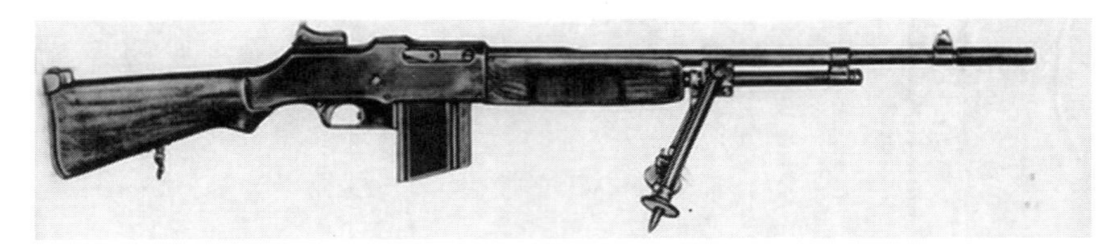

1918A1 BAR

Since the country was involved in no major wars, the weapon was abandoned after being manufactured in small numbers over a two-year period. The A1 saw essentially no action until a few shipments were made to our allies in the early stages of WWII.

In the Hands of Domestic Enemies

As America moved into the automotive age in a big way, those elements of society who elected to earn their livings outside the law found themselves capable of striking swiftly, taking whatever they wanted, killing anyone who interfered, and escaping in a cloud of dust in vehicles whose steel skin in those days was as thick as that of a German helmet. Whereas a .38 Special or .45 ACP round would easily kill a man, it would do little more than dimple Detroit steel.

Outlaws of the day were typically better equipped in both transportation and firepower than the agencies charged with stoppling their violent attacks. Semi-automatic pistols and Thompsons were readily available, along with a vast array of shotguns and revolvers.

Of all the notorious criminals of the period, none stands out more colorfully than Clyde Barrow and his sidekick, Bonnie Parker, both of whom were well acquainted with the automobiles and weapons of the day. Clyde stole a vehicle when he needed or simply wanted one, his preference being a V-8 Ford, which he was so fond of that he wrote Henry Ford a letter praising the car.

Clyde had quite an armory of weapons throughout his criminal career, but the one that he preferred, because of its automatic capability, reliability, and hard-hitting ammunition, was the 1918 BAR.

Where might Clyde hope to find available such firepower, given the fact that most 1918s were in the hands of law enforcement entities or the military? You find such weapons where they are kept, and you take them however you must. Recall that to Clyde Barrow, the whole country was his supermarket, and he never hesitated to requisition whatever he needed at a particular time.

The easiest approach to arming himself in fashion, reasoned Clyde, was to do his weapons shopping in National Guard armories, where BARs and 1911 Colts were kept in great numbers, along with plenty of spare magazines and ammunition, including armor-piercing .30-06 for the BARs. The armories were lightly secured and typically had no after-hours security personnel on hand, so he would do smash-and-grab runs, hitting a number of units. He is known to have raided several armories in the Southwest in Texas and Oklahoma and even in Illinois.

Once equipped with weapons typically in the hands of an infantry squad and with an abundance of ammunition to run those weapons, the Barrow gang became a significant threat to be reckoned with.

Clyde Barrow launched his criminal career at the age of sixteen, when he was arrested for failing to return on time a car he had rented; in a matter of months he was snatched up for stealing some turkeys. His life was downhill from that point on. Over the next few years he robbed stores and cracked safes and stole cars, ending up at Eastham Prison Farm in East Texas in 1930.

Prison life did little to set Clyde's life on the right path. In fact, it was his declared purpose in life after Eastham to wage a "war of liberation" against the prison unit for the abuses he suffered while incarcerated there. The subsequent robbery of stores and gas stations and accumulation of weapons were in preparation for that war.

During his brushes with the law over the next three years, it became quite apparent that Barrow and Bonnie had some serious firepower at their disposal, something considerably more powerful than pistols and Thompsons. The BARs allowed them to survive some epic shootouts in which lawmen (and sometimes even members of the National Guard) vastly outnumbered the gang. Any law enforcement officer with a military background would have recognized immediately the slow-paced hammering of a BAR, of course, but all doubts were removed when in 1933 police raided one of Barrow's abandoned cabins and discovered .45 pistols and a BAR stolen from Oklahoma and Illinois National Guard armories.

Over a period of three years Bonnie and Clyde, often accompanied by others, waged their war against the law, killing several officers during shootouts. In January of 1934 they assaulted Eastham, freeing one of Clyde's old friends, Raymond Hamilton, and several other prisoners, in the process killing a guard. The BAR played a major role in the attack.

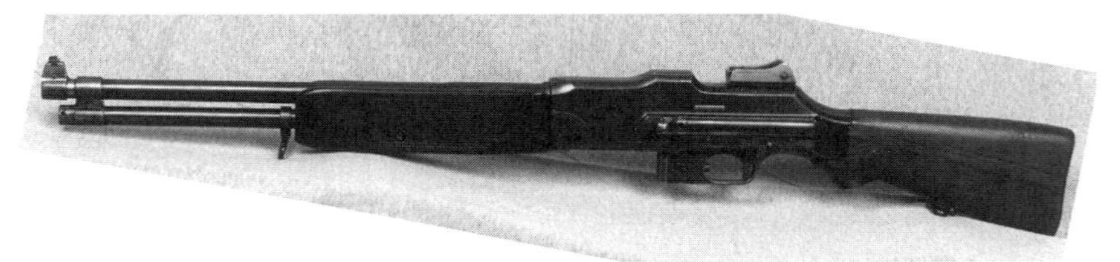

Clyde's "Scattergun" (Courtesy Missouri Highway Patrol Museum)

Armed with several BARs, a couple cut down to make them more manageable in the cars they drove—he called the cut-down BARs his "scatterguns"—Bonnie and Clyde continued their spree, shooting it out with the law whenever necessary, until in May of 1934 they were ambushed on a lonely road in Louisiana. In a hail of bullets from six lawmen, some using BARS, their reign of terror, during which they were responsible for the deaths of nine law officers and a number of civilians, came to an end: 130 bullets struck the bodies of the couple. Inside the bullet-riddled car, among the weapons lawmen found were three BARs. He who had lived by the BAR died by the BAR.

Around a month before he died, Barrow wrote Henry Ford,

praising the V-8s that had helped him to survive so long in a business that "hasn't been strickly legal." Doubtless he might have written a similar letter of praise to John Moses Browning, had Browning not died several years before.

A Marine on a South Pacific Island Stands Vigil with BAR.

The BAR Center-Stage:

World War II and Beyond

The BAR Goes Back to War

Fortunate it was that production of the BAR during the long stretch between the end of World War I and the beginning of World War II never ceased.

It is difficult to determine to what degree our military and political leaders anticipated our eventual involvement in the coming world war, but in April of 1938 the Army made the decision to call for the development of a squad-level light machine gun that could modify existing inventories of 1918s and utilize 1918 spare parts. The 1918A2 was born.

When this country entered World War II, our two heavy machine guns were the water-cooled 1917 water-cooled Browning, firing the .30-06, and the air-cooled M2 Browning, which fired a .50-caliber round. The primary medium machine gun was the .30-06 air-cooled 1919 Browning. All of these weapons could be lugged by one man, but typically the M2 was mounted on a vehicle and operated by one man, and the 1917 and 1919 required at least three soldiers to carry it and ammo and a heavy tripod for stable and sustained firing.

Though in 1937 the 1918A1 appeared on the scene, it did not fully satisfy those in command, but the Army was determined to have a squad-level light machine gun, and nothing in existence could measure up to the legendary BAR for dependability, accuracy, and mobility. Thus, in April of 1938 work began on a significantly modified version of the BAR.

The 1918A2 featured an easily removed barrel-mounted bipod and a hinged buttplate, but instead of a choice between semi-auto and full-auto fire, the operator could now select between two automatic rates of fire: 350 rounds per minute or 650 rounds per minute. Further, the rifle was equipped with a magazine guide, which facilitated faster replacement of a spent magazine, and a shortened handguard with heat shield. Finally, the weapon was fitted with a new flash suppressor and fully adjustable rear sight. Later, a removable handle was provided, permitting easier carrying of the gun, especially when the barrel had heated up after extended firing.

World War II: The BAR at Work

When World War II erupted, the pace of production of 1918A2s was quickly increased, though the steadily dwindling availability of 1918 parts and the lack of serviceable tooling to make more resulted in new production, which fell initially to IBM and New England Small Arms. Before the war was over, these two companies had produced over 200,000 1918A2 BARs.

As the war continued, demand for the BAR in the European and South Pacific theaters put increasing pressure on manufacturers. In 1942, a shortage of black walnut for buttstocks resulted in the introduction of the synthetic Bakelite stock, which was provided by Firestone Rubber and Latex Products.

Production of the 1918A2 greatly increased in 1943, when IBM introduced a new method of building receivers, which to that point had to be machined from pieces of solid forged steel, involving much time and waste of material. The new method, perfected by General Motors in manufacturing automotive components, involved the use of ArmaSteel, a type of cast iron made from scrap steel and pig iron that were heated and blended and cast to form receivers that could then quickly be machined to their final configuration. These new receivers were thoroughly tested and determined to be as strong and resistant to wear as the earlier forged-steel versions.

1918A2 BAR

During World War II the Army infantry squad consisted of a dozen soldiers, with one BAR and eleven M1 Garands, with at least one of the rifle bearers assigned the role of carrying extra bandoliers of magazines for the BAR.

When the Japanese attacked Pearl Harbor, the Marine squad numbered nine, with one BAR and eight riflemen, but within a month after America's declaration of war, the squad was increased to thirteen, still with only one BAR. As the war progressed in the Pacific, the Marines once again adjusted the size and firepower of their squads by reducing the number of men to twelve but assigning two BARs to each squad.

The effectiveness of the BAR in the Pacific theater led to another adjustment in squad size and strength when in 1944 the Marines restructured the squad, with a squad leader and three four-man fire teams, each with a BAR man and an assistant BAR man. Now the squad had significantly increased firepower, with three BARs and a larger ammo supply for each.

The BAR was a highly effective weapon that proved its merit time after time as our troops fought the Japanese foot by foot across the islands, serving to suppress enemy fire, disrupt human-wave attacks, and function as an effective weapon against snipers and enemy troops firing from dugouts and tunnels. Most Marines regarded the BAR as a highly reliable and accurate weapon, and anecdotes abound that attest to its role in our winning the war against the Japanese. The BAR man often carried the day.

Whereas it is true that the BAR was the darling of the Marines and Army as they took island after island in the Pacific, the weapon likewise proved its worth in the European theater, beginning with the Normandy Invasion.

As our troops moved steadily toward Berlin, the BAR was there, reliable as always under the most adverse of conditions and providing our infantry squads the portable automatic weapons so desperately needed against the heavily armed German units.

Again, one can find story after story of the exemplary role of the BAR as our forces battled the Germans to their ultimate defeat.

Variants of the BAR were developed by a number of other nations during the war, including Belgium, Poland, and Sweden, and China used both Belgian FN1930s (BAR variant) and American-made 1918s against the Japanese. By the end of the war, the BAR had earned universal respect as a durable and deadly light machine gun.

The BAR at War: Korea and Beyond

Before World War II was over, John Browning's BAR had established a reputation for itself all across the globe, serving as an issued arm in all branches of service in the American military and in the armies of a number of other countries.

As this country settled into an uneasy peace with the Russian empire and its vast industrial military complex, the prevalent notion residing among those in charge in Washington seemed to be that because of the introduction of nuclear weapons into the equation, the era of large ground wars was over. All interest in developing new conventional arms diminished as our troops adjusted to garrison duties in Europe and Asia or returned to the work force at home.

When hostilities broke out on the Korean Peninsula in June of 1950, the weapons used by both United Nations and North Korean/Chinese forces were almost exclusively those left over from the recent War. Most of our troops carried the semi-automatic M1 Garand as their primary weapon, with M2 carbines, Thompsons, and M3 "Grease Guns" for close-in automatic support. Browning 1917 and 1919 .30-caliber machine guns and M2 .50-cals addressed the needs for longer-range automatic fire.

As was the case in both major theaters of World War II, the 1918A2 BAR was assigned for use at the squad level, giving our troops the mobility that weapon provided, plus the capability for short-range and long-range applications. It could lay down suppressive fire in automatic mode on the immediate flanks or against frontal assault or take out a sniper at 500 yards with its powerful 30-06 round.

Tactics employed by enemy forces during the war made it difficult for established machine gun positions, using the heavier Brownings, to effectively protect our troops as they advanced across hostile terrain. The BAR man, though, could quickly maneuver his weapon to address whatever threat was at hand. So great was its flexibility that the BAR frequently became a brigade-level weapon, shunted about to defend major perimeters from swift flank assaults characteristic of the Communist forces they were engaged with.

In short, the very attributes of the weapon that made BAR so effective in World War II assured its popularity in the Korean War: It was highly mobile, extremely reliable, and inherently accurate, and it could use ammunition stripped from machine gun belts or M-1 clips or left lying about in ammo cans or as loose rounds.

As was always the case in WWII, there was no shortage of soldiers eager to serve as BAR man or ammunition assistant. Anecdotal evidence abounds that the BAR was one of the most effective weapons of the war, time after time demonstrating its ability to perform reliably under the most adverse conditions imaginable: mud, sand, ice, snow. Nothing prevented the BAR from hammering away at the enemy.

Early on in the Korean conflict, the BARs in service suffered a higher than usual frequency of mechanical issues, but the cause was quickly isolated and addressed: The weapons that had problems were almost without exception those that had undergone armory reconditioning during which over-used recoil springs were not replaced. Once the recoil springs were replaced, the BARs performed like new.

Demand for the 1918A2 during the Korean War was such that an additional 61,000 were produced during the period by Royal Typewriter of Hartford, Connecticut, who kept costs down by carrying on the tradition established by manufacturers of the weapon in the latter stages of World War II of making cast receivers and trigger housings of Armasteel.

The 1918A2 BAR, then, carried on its fine tradition of mobility, reliability, and accuracy throughout the Korean War, after which it was once again relegated to training facilities or distributed to National Guard armories to await its ultimate fate.

During the early stages of the Vietnamese War the BAR was issued to the South Vietnamese Army and its regional allies, where it rose to the occasion and often was the weapon of choice for those who could lay hands on it. It was also used by our Special Forces during the war. James Ballou, whose *Rock in a Hard Place* is the most comprehensive study of the BAR ever published, writes that during his three tours in Vietnam, he "thanked God for . . . having a BAR that actually worked, as opposed to the jamming M16 . . ." (Ballou 204).

Few weapons that have ever appeared in the U.S. arsenal have served as long and well as John Moses Browning's iconic BAR.

After the failure of the M14 to replace the BAR as a squad-level light machine gun, our military was essentially without such a weapon until the introduction of the M249 Squad Automatic Weapon (SAW) in 1984, which, though belt-fed, could be fired from the shoulder. The M249 weighed in at roughly the heft of the BAR, and it was capable of a high rate of fire (up to 800 rounds per minute), but the much smaller 5.56 round had nothing like the knock-down, penetrating power of the .30-06.

The BAR remained in National Guard armories into the 1970s, and several countries—Belgium, Portugal, and Spain among them—kept the weapon in service until almost the end of the century.

Light machine guns will come and go over the years, but none is likely ever to enjoy the enduring romantic allure of the BAR.

III

EVERY MAN'S BAR:

THE OHIO ORDNANCE 1918A3-SLR

OOW 1918A3-SLR with Bakelite Buttstock

OOW 1918A3-SLR with Walnut Buttstock

An Old Warrior Comes to Life: The Ohio Ordnance BAR 1918A3-SLR

Bob Landies: The Man Behind the Gun

Bob Landies admits that he had a "gun problem" at a very early age. He has a picture of himself at age five or so holding a rifle in one hand and a pistol in the other, with a German helmet on his head. His grandfather was an infantrymen in World War I, and his father slogged ashore at Normandy and engaged in a number of battles in World War II, so Bob had military weapons in his background. Though his father was not a serious gun collector, he was nonetheless a gun enthusiast, and that enthusiasm for guns rubbed off on young Landies. Over the years Bob Landies' enthusiasm for guns continued to grow.*

Bob did not encounter his first BAR until he joined an Army National Guard unit in Ashtabula, Ohio, in 1964. As a member of a recon section of a tank battalion, he was issued a 1918A2, which he found far more gratifying than the M-1, with its eight-round clip.

---

*All comments by Bob and Robert Landies, whether quoted or summarized, are based on phone interviews with Paul Ruffin In November of 2014.

Shortly after his graduation from high school, Bob went to work for Thompson Ramo Wooldridge (TRW) in Cleveland in the Navy division, helping build Mark-46 torpedoes. The company was also manufacturing M-14s. The plant work force was, as might be expected, largely veterans of WWII and the Korean War, so guns were frequently subjects of conversation.

Exposure to these veterans and their experiences with military weapons added impetus to Bob's interest in guns. His work required a certain measure of production in a day, so he would often work through early breaks and lunch and take off the last couple of hours to search for guns and other types of militaria.

It is not surprising that during this period he was reading everything he could find on John Moses Browning. "He was anointed," Bob says. "He had a gift. A dispensation. And the more you read about him . . . I tell ya what amazes me: I worked so hard to have one design proven out or one engineering thing come to fruition, ya know, and the things that guy did and the short period of time that he achieved them is the most astonishing part for me."

When Bob left TRW, he worked in the construction business for ten years, where he learned not simply the construction end of things but also the financial. When the company decided to relocate him to Texas, he left that job. His response to them was, "I don't speak Spanish, I hate refried beans. I love Texas, but I don't wanna live there the rest of my life."

The real break in Bob Landies' professional life came soon after. He bought two pistols from a friend—a Luger carbine and a broomhandle—and took them to the Ohio Gun Collectors' Show in 1981. He did not take the guns to sell, merely to find out more about them from dealers and buyers, but when a man asked to see the guns at the door and offered Bob $10,000 for them, he took the money and his first step toward a career in guns.

Over the next couple of years he haunted the gun shows, buying and selling and doing repairs on weapons. He bought a lot of registered full-auto BAR receivers from an industry in Kentucky and started building the machine guns from parts sets, which were in abundance in those days. Sales were brisk, and Bob would make up to $500 profit on each gun. When the Gun Control Act of 1986 went into effect, making the manufacture of machine guns and limiting their sales to law enforcement and the military, Bob's BAR production was over.

The Birth of the 1918A3-SLR

A question that Bob heard a number of times during his days building the BAR and after that particular phase of production ceased was, "Do you have a semi-automatic version of the BAR?" His answer was always the same: "No. Nobody does. They don't exist." In time, though, that question engendered in his mind a notion: "Why don't *I* design and build a semi-automatic version of the BAR?"

Bob never was a design engineer, but he had general ideas on how something should be built, so he enlisted the help of Jim Taylor, out of Pennsylvania. He explained his concept for the semi-automatic BAR, Jim came up with a design for a prototype, and they collaborated on building it.

The problem with their first production design was that it was too easily converted to full-auto, so the ATF rejected it. For the next year and a half Bob dedicated himself to coming up with a version of his BAR that would retain as much of the appearance and function of the 1918A2 as possible and at the same time meet the requirements imposed by the ATF.

So, Bob explained:

> *I spent a year and a half refining the whole thing to make it fully acceptable from the closed bolt to ATF and reliably self-loading. I like to use that term,* self-loading, *because* semi *is like* semi-pregnant. *I mean, either it is or it's not. So I used that term* self-loading rifle. *It's a British term, the self-loader, but it's very applicable. You fire the first round and it self-loads the second one.*
>
> *So I dubbed it the 1918A3 Self-Loading Rifle (SLR). California didn't catch on, New York didn't catch on. I just kept making rifles and selling'm. That's kinda an overview. Jim Taylor and I worked together and evolved the system and also, I*

had two really heads-up tool makers in my shop so it was a cooperative process.

I can't say that I developed . . . uh, the idea was mine and it was predicated on a customer demand and I said, "Wait a minute, ya know, here, let's identify a need." The need is, I need semi-auto rifles. Wow. For us to jump, um, from repair work to manufacturing is a pretty big jump.

All this while, Bob was continuing with his regular shop work and facing mounting financial burdens and wondering just how he was going to keep the BAR project rolling.

His salvation came in the form of a request from a friend of his for a batch of Browning M2s, which Bob had been for some time engaged in repairing and restoring. Within sixty days he ended up supplying the man with 150 M2s, procured as complete guns and parts sets from numerous sources, which provided him with the cash to get on with his pet project, the 1918A3-SLR.

One of the initial problems with the SLR project lay in the receivers, which they had to make themselves. A real break came when in the aftermath of the 1986 gun laws, which prohibited the manufacture of machine guns for sales to individuals, a friend of Bob's sold him a receiver mold that he had been using for full-auto BARs. Bob modified the mold to facilitate the production of the SLR receiver, using investment-cast 8620 steel that is precisely machined using CNC technology, and over the years has built his own molds, continuing to fine-tune them as new ideas develop to improve the design.

The first copies of the 1918 A3-SLR were produced under the name Collector's Corner. Bob's shop was in the corner of his basement, and he was a collector—hence the name.

As interest in the new BAR continued to build, thanks in part to a number of World War II movies that came out during that period featuring the BAR, it was obvious that Bob was going to have to expand his operations to accommodate demand, so he incorporated under the name of Ohio Ordnance Works and began scouring the country for BAR parts kits.

Another big break came in 1996, when a friend of his told him about a lot of 3000 1918A2s in Belgium that had been initially been shipped to Jordan under a lend/lease agreement. All were reconditioned to excellent condition or new, and Bob managed to buy them for $175 each, but he also had to send a team of workers over to break the weapons down and ship them back to the States. They arrived just days before the Clinton administration introduced their initiative to ban the importation of firearms.

Today Ohio Ordnance Works continues to build the highly successful 1918A3 BAR, using mostly newly-manufactured parts. Bob refuses to use USGI parts that do not meet his exacting standards. When a customer buys an Ohio Ordnance BAR 1918A3-SLR, the gun is new from tip to tip and from the outside easily passes as an original 1918A2 (but for the OOW identification stamping on the receiver). The weapon fires from a closed-bolt system and utilizes a fire-control group designed for the special self-loading receiver, but it utilizes any USGI magazine and to a very large degree functions like an original 1918. It is a finely made machine that John Moses Browning himself would be proud to claim.

The hefty A3 (19.4 pounds with unloaded magazines) comes in a standard package or as what OOW refers to as their 1918A3 Bundle. The beautifully parkerized weapon is available with either an American walnut or Bakelite buttstock. The company also has available a number of spare parts and accessories for the weapon.

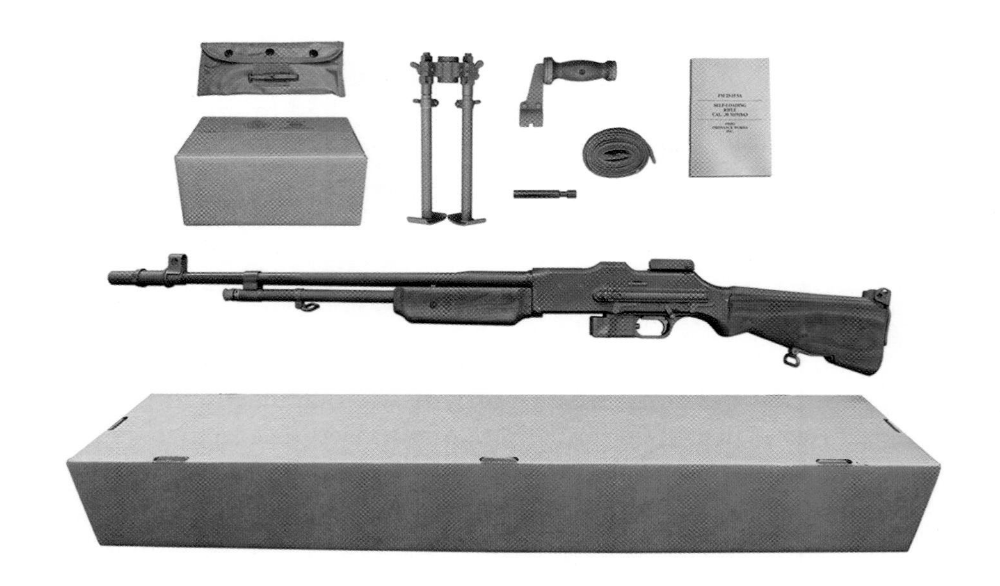

OOW 1918A3-SLR Basic Package

BASIC PACKAGE INCLUDES:

•(2) 20rd. Magazines

•Web Sling

•Bipod

•Carrying Handle

•Flash Hider

•Bolt Hold Open Device

•Cleaning Kit

•Manual

OOW 1918A3-SLR Bundle

BUNDLE INCLUDES:

• Walnut or Bakelite 1918A3-SLR
•(2) 20rd. Magazines*
•Web Sling
•Bipod
•Carrying Handle
•Flash Hider
•Bolt Hold Open Device
•Cleaning Kit
•Manual
•Custom Cut Pelican Case
•1918A3-SLR Maintenance Kit
•(2) 30rd. Magazines
•Membership Site Access for Video Tutorials
•Certificate for Inspections and Re-Park
•Field Stripping and Cleaning DVD

1918A3-SLR in Pelican Case

1918A3-SLR Cleaning Kit

1918A3-SLR Maintenance Kit

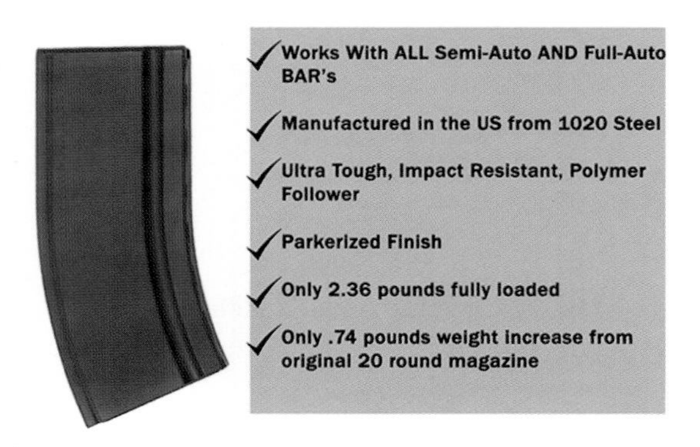

New 30-Round Magazines for BAR

Heavy Counter-Assault Rifle (HCAR)

The BAR Goes Mod: The HCAR

Robert Landies and the HCAR

No good gunmaker ever gives up the passion for perfecting whatever design he has come up with, and such was the case with Bob Landies and Ohio Ordnance Works.

There was no questioning the success of the 1918A3-SLR, which was enthusiastically welcomed by American shooters and collectors who longed to possess a BAR but could not afford "the real thing." They discovered that the OOW A3 might not fire 350 to 550 a minute with a single trigger pull, but everything else about the weapon declared it BAR all the way.

The wheels were still spinning in Bob's mind: *How can we make it lighter, improve the cooling, modernize our BAR, and keep it a BAR*? The answer was simple: *Just do it!*

And the Heavy Counter Assault Rifle was born, though Bob gives credit for the idea, design, and development of the HCAR to his son, Robert.

Under the leadership of Robert Landies, the HCAR became a production weapon weighing over seven pounds less than the A3, featuring a redesigned selector switch, incorporating a bolt hold-open function, and utilizing a thirty-round magazine. Upon first glance, though, it is clear that this gun, essentially identical in many respects to the A3 on the inside, is a weapon for the 21st Century.

Through relief machining, the receiver was lightened considerably, and by dimple-machining the barrel, OOW both lightened the gun and aided cooling by providing more surface area on the barrel. The BAR was always recognized as an "overbuilt" weapon, with massive parts, so the relief machining of the forged receiver and the barrel in no way compromised the strength of the gun.

Robert explained how the HCAR project came to be:

> *Well, we use an investment-casting process to produce the receivers for the 1918A3, as you know, and from time to time there would be little blemishes on the outside surface—nothing that in any way would have weakened the receiver, but we wanted as perfect a surface as possible—and those pieces got kicked off to the side, orange-tagged, not because they were scrap—nothing wrong with them mechanically or functionally, just an aesthetic issue—but they were not usable in the A3 building process.*
>
> *These receivers started accumulating, maybe a stack of twenty or so, and I got to wondering what we might do to take advantage of them.*

Robert decided to design an entirely different version of the BAR, one that, among other things, would be lighter, utilize a thirty-round magazine, and be equipped with rails that would allow it to be dressed out with a wide range of accessories.

Part of the lightening process involved machining away the outside surfaces of the receiver, maintaining the same degree of strength and durability expected of the A3 but reducing the weight by over a pound, and using lightening dimples on the barrel instead of taking the more conventional approach of fluting it. He explained:

> *I've always looked at fluting from an engineering standpoint as dumb. It just didn't make any sense because you're putting a low point, or a stress point, in the barrel. And then you're taking that and you're raking it all the way down the barrel. So*

Receiver of HCAR Demonstrating Relief Machining

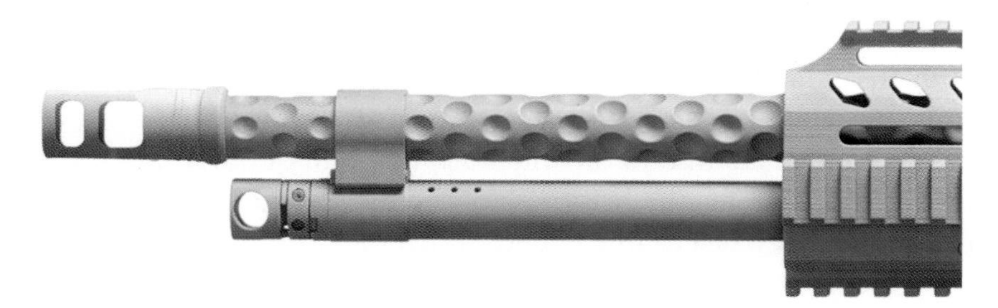
Dimpled Relief Machining on HCAR Barrel

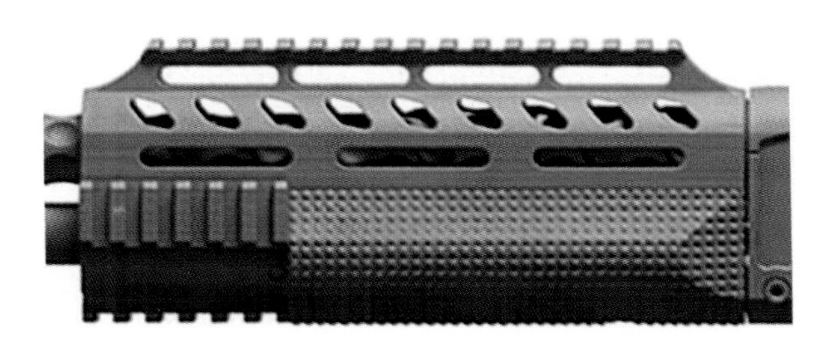
HCAR Handguard

you're putting a low point the whole distance. There's no material in between, you know, unlike the dimples, where there's material in between each one.

They made some major changes to the handguard, creating a U-channel, permitting full hand coverage, with a rail on top. With the front and rear rails, the HCAR can handle almost any accessory that the AR platform supports.

Although the internals of the weapon are not vastly different in operation from those of the A3, Robert did redesign the firing pin so that it is captured in the bolt and will not accidentally fall out. He also developed an entirely different trigger group and incorporated a bolt hold-open device that locks the bolt back when the last round is fired from the magazine. Further, he moved the magazine release from inside the trigger guard to a position just behind the magazine, so that, as with an AK, the lever can be operated with either hand. One final refinement in the trigger group was a redesign of the selector switch: "I came up with a different selector and it's a short, with a 45-degree throw, so it's not very hard to move."

One outstanding feature of the new trigger-group design is that the HCAR has a single-stage trigger, which, in conjunction with the new disconnecter and sear and spring, has what Robert calls "a really nice short pull and a short reset," resulting in a five-pound pull, compared to the much heavier pull of the A3. In a video he demonstrates the briskness of the new trigger by rattling off rounds almost a fast as one would expect of a fully automatic BAR.

Another improvement that Robert wanted on the HCAR was a redesign of the charging handle, which he felt needed to be able to accommodate two fingers and yet not interfere with the operation of the weapon by hanging on such things as web gear.

So I wanted something flat and streamlined so it wouldn't catch, and then I was thinking, well, if I'm gonna do something, first of all, I'd want to get two fingers on it. Second of all, I want it to fold down so it doesn't catch on anything, so both those objectives are accomplished. It's almost two times as long, you can definitely get two fingers on it, and it folds down to less than half of the original height.

HCAR Package

Robert also redesigned the buffer system, which functions with a new muzzle-brake/compensator to reduce recoil and help suppress muzzle rise. “Overall,” he says, “the gun shoots just how I wanted it to. It is a blast to shoot, and everyone loves the way it handles.”

HCAR in Black

HCAR in Flat Dark Earth

HCAR in OD Green

The features of the HCAR (from the OOW catalog):

It fires the same power-packed .30-06 round as the original BAR. I'm sure you already know that out of the box, the standard .30-06 round is not only deadly accurate up-close, but also packs more punch at greater distance than the standard .308 round can—so the .30-06 caliber gives you the best of both worlds.

The rifle's built on a rugged, investment cast receiver made of 8620 steel, relief cut for weight reduction. And the sturdy receiver handles any of today's quality high-pressure loads without missing a beat, giving you a broad range of ammunition options so you can fine-tune your rifle for your specific application.

The H.C.A.R.'s action is proven under fire: The rifle fires from the closed bolt using the same patented action as our semi-auto BAR (the 1918A3), that's been field tested and verified by thousands of shooters over the past 20 years . . . .

The weight has been drastically reduced—it's over 7 pounds lighter than the original BAR—which means faster, easier handling and more comfortable carrying and shooting.

The receiver's integrated picatinny rail allows you to set up a mounted optic, as well as a back-up sight. The H.C.A.R.'s barrel has been dimpled using a patented design which actually increases the barrel's surface area yet at the same time reduces weight, maintains rigidity and strength, and accelerates barrel cooling.

While the rifle comes standard with a Surefire muzzle brake, the barrel is threaded with the industry standard 5/8-24 thread to allow you to add whatever flash suppressor or muzzle brake (including those used for suppressors) you choose.

The rifle comes with a newly designed hand-guard with four integrated picatinny rails on it. These rails allow another optic device set-up (think thermal with your main optic, etc.), as well as the attachment of your choice of bipods, sling swivels, lights, etc. Not only that, but the trigger housing has been completely redesigned. From the outside, it includes an easier to engage safety with a shorter throw.

Internally, it features a newly designed trigger with a crisp, factory-set, 6 pound trigger pull, a short travel when firing, and an even shorter trigger reset, giving you the ability to fire your rifle at near full-auto speeds.

The H.C.A.R.'s newly designed, ergonomic pistol grip allows for a more natural wrist alignment and enhanced shooting comfort.

The easy to reach, intuitive mag release paddle is quickly engaged in the same fluid motion as unloading and reloading the magazine, so your time between mag changes is lightning fast.

An integrated bolt hold-open mechanism locks the bolt to the rear when your magazine is empty OR whenever you manually pull the bolt back without a magazine in the rifle—this allows single feeding, simple "one look" check to ensure a clear chamber, and easy field cleaning.

A quick flick of the ambidextrous bolt release returns the bolt back to the firing position. The release is conveniently located just forward of the trigger guard, easily reachable with the thumb of your mag reloading hand. This feature makes reloading a fresh magazine and releasing the bolt possible in one fluid motion with the same hand, increasing your speed during the reloading cycle.

The felt recoil of the rifle has been sliced in half by upgrading to a hydraulic buffer system from the original cone buffering system.

The internal threads of the buffer housing allow you to attach either the mil-spec or aftermarket AR15/M4 butt-stock/buffer tubes. That means you can add your choice of any AR15/ M4 butt-stock you want to your rifle.

While the original BAR's smallish magazine capacity was a knock on the gun, we've increased the H.C.A.R.'s magazine capacity by a full 50%, giving you 30 rounds per magazine vs. the measly 20 round mags of the original BAR. This way you'll bring more fire-power to the party, cut down on loading and unloading cycles, and stay in the action longer.

The HCAR was initially offered in four different finishes—Black, Coyote Tan, Flat Dark Earth, and OD Green—though because the Coyote Tan and Flat Dark Earth were so similar in color, the former was dropped from the line.

1918 SLR

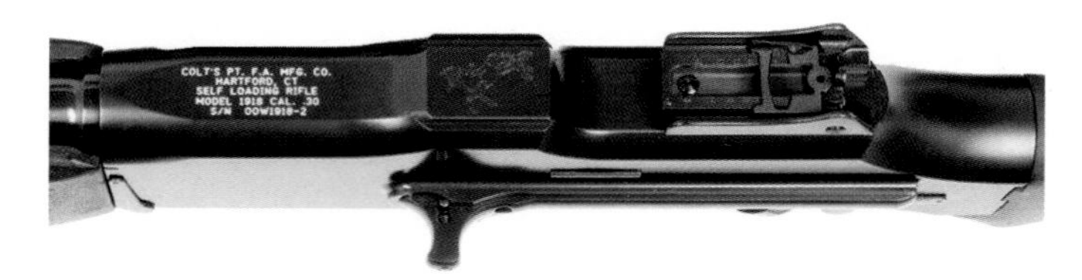

Top of Receiver, 1918 SLR

Resurrecting the 1918

The New 1918 SLR

Not content to curb his momentum, Bob set to work designing an exact reproduction of the original 1918 BAR, at least from the outside. Though for safety and manufacturing purposes OOW had to make some internal upgrades, externally the gun is identical to the 1918. Bob said that you could lay his new 1918 SLR alongside an original Winchester or Colt, and you would not be able to tell the difference. Like the original, the 1918 SLR will come with neither bipod nor carrying handle.

Bob pointed out tht this new 1918 Self Loading Rifle will actually bear the Colt logo:

> *We just did a contractual agreement with Colt and we're going to build a blued 1918 self-loading rifle. They're going to have the Colt name on them, the Colt logo, and they're going to be in a specially designed and manufactured in Italy leather carrying case.*

Bob's enthusiasm for his new baby is obvious when he talks about the rich blueing and the furniture: "It has the dual handguard that wraps around. It's got beautiful wood—you ought to see the walnut. We've got a manufacturer in Pennsylvania who's doing the walnut. I mean, it just knocks your socks off." The receiver on the 1918 SLR is machined from billet so that the blueing is deep and rich and matches the blueing on the barrel, yet another extra demonstration of OOW to be as authentic as possible with the weapon.

There is little doubt that this newest addition o the OOW BAR lineup will be a grand success. The first production run will be for 1000 weapons, and already orders are waiting to be filled.

The Future for the BAR

Whatever the future holds for the Browning Automatic Rifle, Ohio Ordnance Works will be right in the middle of it. Bob Landies' enthusiasm and enormous talent appear to have been passed along to Robert, so OOW will continue to bear the standard for this fine weapon. Its rich history will never be forgotten, thanks to the effort of this small company in Ohio to keep John Moses Browning's fine creation alive.

Selective Bibliography

Arrington, Leonard. “Browning, John Moses (1855-1926).” *The New Encyclopedia of the American West*. New Haven: Yale University, 1998.

Ballou, James L. *Rock in a Hard Place: The Browning Automatic Rifle*. Cobourg, Ontario: Collector Grade Publications Inc., 2000.

Barrow, Blanche Caldwell and John Neal Phillips. *My Life with Bonnie and Clyde*. Norman: University of Oklahoma Press, 2004.

Bishop, Chris. *The Encyclopedia of Weapons of World War II*. New York: Sterling Publishing, 2002.

Browning, John, and Curt Gentry. *John M. Browning, American Gunmaker: An Illustrated Biography of the Man and His Guns*. New York: Doubleday, 1964.

Burrough, Bryan. *Public Enemies*. New York: The Penguin Press, 2004.

Canfield, Bruce N. *U.S. Infantry Weapons of the First World War*. Lincoln, RI: Mowbray Publishers, 2000.

__________. *U.S. Infantry Weapons of World War II*. Lincoln, RI: Mowbray Publishers, 1998.

Chinn, George M. *The Machine Gun, Volume I: History, Evolution, and Development of Manual, Automatic, and Airborne Repeating Weapons*. Washington, D.C.: Bureau of Ordnance, Department of the Navy, 1951.

George, Lt. Col. John. *Shots Fired In Anger*. Fairfax, VA: NRA Publications,1981.

Goldsmith, Dolf L. *The Browning Machine Gun: Rifle Caliber Brownings in U.S. Service, Volume I*. Cobourg, Ontario: Collector Grade Publications, 2005.

__________. *The Browning Machine Gun: Rifle Caliber Brownings in U.S. Service, Volume II*. Cobourg, Ontario: Collector Grade Publications, 2006.

__________. *The Browning Machine Gun: Rifle Caliber Brownings in U.S. Service, Volume III*. Cobourg, Ontario: Collector Grade Publications, 2008.

Guinn, Jeff. *Go Down Together: The True, Untold Story of Bonnie and Clyde*. New York: Simon & Schuster, 2009.

Hodges Jr., Robert R. *The Browning Automatic Rifle*. New York: Osprey, 2012.

Hogg, Ian V. *The American Arsenal*. Mechanicsburg, PA: Stackpole Books,1996.

___________. *Machine Guns: 14th Century to Present*. Iola, WI: Krause Publications, 2002.

___________ and John Weeks. *Military Small Arms of the 20th Century*, Seventh Edition. Iola, WI: Krause Publications, 2000.

Iannamico, Frank. *Hard Rain: History of the Browning Machine Guns*. Harmony, MA: Moose Lake Publications, 2002.

J.M. and M.S. Browning Company. *A History of Browning Guns from 1831*. St. Louis: Browning Arms, 1942.

Knight, James R. and Jonathan Davis. *Bonnie and Clyde: A Twenty-First-Century Update*. Austin, TX: Eakin Press, 2003.

Kokalis, Peter G. "Ohio Ordnance Works' HCAR," *Shotgun News* (November 20, 2014): 4-5, 9, 12-13, 16-17, 20, 22.

Marshall, S.L.A. *Commentary on Infantry Operations and Weapons Usage in Korea, Winter of 1950-1951*. Chevy Chase: Johns Hopkins University Press, 1951.

Miller, David. *The History of Browning Firearms*. New York: Chartwell Books, 2014.

Milner, E.R. *The Lives and Times of Bonnie and Clyde*. Carbondale: Southern Illinois University Press, 1996.

Parker, Emma Krause, Nell Barrow Cowan, and Jan I. Fortune. *The True Story of Bonnie and Clyde*. New York: New American Library, 1968. (Originally published in 1934 as *Fugitives*.)

Phillips, John Neal. *Running with Bonnie and Clyde, the Ten Fast Years of Ralph Fults*. Norman: University of Oklahoma Press, 1996, 2002.

Prendergast, Gareth J. "Scaring Them to Death: The BAR." *World War II* 21.3 (2006): 18-20.

Ramsey, Winston G., ed. *On The Trail of Bonnie and Clyde*. London: After the Battle Books, 2003.

Reid, Kevin B. "John Moses Browning." In *Great Lives from History: Inventors and Inventions*. Vol. 1. Ed. Alvin K. Benson. Hackensack: Salem, 2010.

Rottman, Gordon L. *Browning .30-Caliber Machine Guns*. New York: Osprey Publishing, 2014.

Smith, W.H.B. and Joseph Smith. *Small Arms of the World*. New York: Galahad Books, 1973.

Steele, Phillip, and Marie Barrow Scoma. *The Family Story of Bonnie and Clyde*. Gretna, LA: Pelican Publishing Company, 2000.

Treherne, John. *The Strange History of Bonnie and Clyde*. New York: Stein and Day, 1984.

Utter, Glenn H. and Robert J. Spitzer. "Browning, John Moses (1855–1926)." *Encyclopedia of Gun Control and Gun Rights*. 2nd ed. New York: Grey House, 2011.

Webb, Walter Prescott. *The Texas Rangers: A Century of Frontier Defense*. Austin: University of Texas Press, 1935.